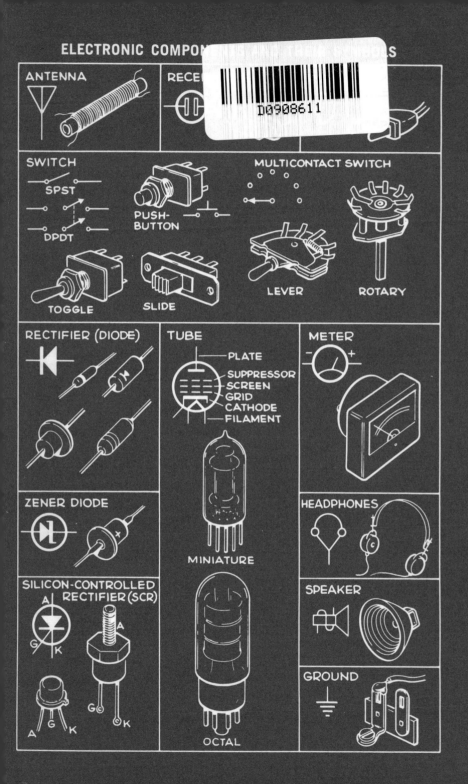

ELECTRONICS
for EVERYBODY

ELECTRONICS
for EVERYBODY

RONALD BENREY

Drawings by
Richard Meyer

Popular Science
Publishing Company

Harper & Row
New York, London

Library of Congress Catalog Card Number: 74-115430

Designed by Jeff Fitschen

Manufactured in the United States of America

TO JANET

Contents

ELECTRONICS
for EVERYBODY

How You Can Get Started with Electronics

Electronics paints pictures across your TV screen, pumps sound through your stereo speakers, sets the shutter speed or f/stop of your automatic camera, and controls the speed of power tools in your workshop—to list just a few of its accomplishments. Name your own special interest or hobby, and the chances are good that at least one electronic device helps you enjoy it.

This is the reason why, each year, thousands of people take soldering irons in hand for the first time, and begin to work with electronics. Some are electronic hobbyists—folks who enjoy building electronic gadgets because they are intrigued by the science and technology of electronics. Their special interest *is* electronics. Many more, though, are drawn to electronics because of some other interest. An audiophile assembles a stereo amplifier from a ready-to-wire kit, for example. Or a car buff builds an electronic tachometer to help him tune engines. Or an amateur photographer builds an electronic timer for his enlarger. To these people, electronics represents a means to an end. They are willing to work with electronics to save money over the cost of a commercially manufactured electronic device, or to build a gadget they can't buy, or to expand their involvement in their main hobby.

These people find enjoyment in this work. Electronics is a thoroughly fascinating subject, and building electronic devices and projects is, in many ways, a very rewarding pastime. The areas you can work in, and the types of projects you can build, are almost unlimited—amateur radio, audio reproduction, remote control for models, automotive accessories, electronic musical instruments—yet to get started you'll require only a modest investment in a few basic tools and simple test instruments.

Building projects is an excellent way to learn about electronic technology. And you'll find that you don't have to understand electronic theory to build most projects—even from scratch. Even though you may work from a published diagram or circuit layout plan in a magazine or book, when you build from scratch, you'll find numerous occasions to add your own creative touches to a project, especially in the selection of construction techniques, control-panel layout, and overall appearance.

But how and where do you begin? In many years as both a designer of electronic projects and technical editor concerned with electronic subjects, I have received scores of letters from readers all asking the same question: "How do I start working with electronics?"

The beginner can get started in electronics by building simple kits to make useful projects, such as intercom systems, burglar or fire alarms, simple radios, electronic "eyes," and many others, all available at electronic and craft stores.

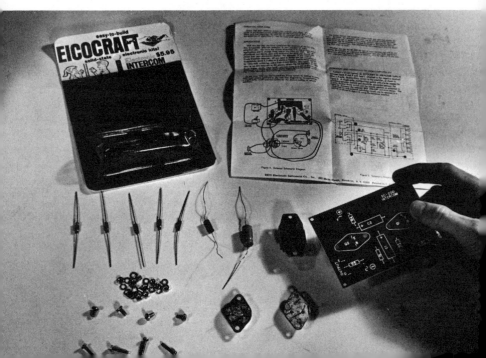

With no experience at all, the hobbyist can wade into a kit for a fine sound system, or one of hundreds of other electronic projects, by following the simple step-by-step instructions from start to finish. Here the parts of a Heathkit stereo compact sound system are laid out prior to assembly. The walnut outer case remains unpacked in another box.

Many kits that come with prewired and assembled sections are available to the home-electronics builder. These generally cost more, but save the hobbyist the trouble of soldering resistors, capacitors, and other tiny components into printed circuit boards. He need only wire the sections into the chassis in the final construction, following step-by-step instructions in the well-prepared and easy-to-read manual.

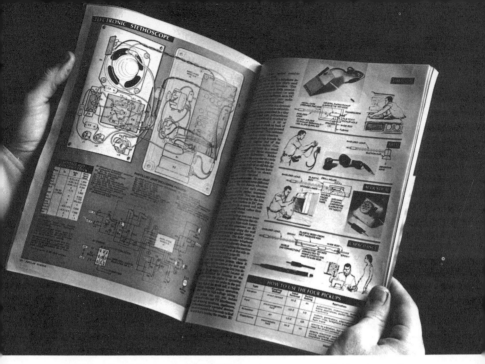

The budding electronics hobbyist can turn to many electronics and general how-to magazines for projects to build. Carefully planned text and layouts, as in this *Popular Science* article, give a photo of the finished product, a pictorial showing the exact position of each component and its wiring, and a schematic diagram which is easy to read in relation to the pictorial. The text generally explains how the project works and gives construction procedures, and a parts list completes the article.

It's paradoxical that the aura of mystery and technical wizardry that surrounds electronic devices can simultaneously attract and frighten potential hobbyists. There's something about electronics that makes many people shy away. Perhaps it's the fact that electronic devices work by harnessing fast-moving streams of tiny bits of electric charge called *electrons*. Because we can't see electrons, it may seem incredible that we can design and build devices that capitalize on their motion, and what seems incredible usually seems difficult.

Or perhaps it's the strange appearance of the many different components used to build electronic circuits—buglike transistors, candy-cane striped resistors, and bullet-shaped diodes, to name a few. It also may be that some people think that soldering wires together is a complicated, hard-to-learn skill, when in fact it's really an easy technique to master. Whatever the reasons for this public ambivalence toward electronics, I hope this book will help to resolve it. For I have included the answers to most of the questions that newcomers to electronics invariably ask.

But keep this point in mind as you read: electronics is a curious mixture of technology and craftsmanship and it is difficult to approach the subject from either the purely theoretical or the purely practical point of view. It's therefore advisable that when you read about electronics you should work with electronics at the same time.

The Ways You Can Work with Electronics

Some people build projects from scratch; others assemble ready-to-wire kits. Which do you fancy? Before you answer, though, take a closer look, for buried inside each of these pleasant pastimes is a wealth of activities.

Want to build projects from scratch? You'll find ideas, inspiration, and circuit diagrams in scores of articles published each month in hobby and technical magazines. Some describe simple projects; others talk about electronic behemoths that rival the electronic systems in a space capsule. Many of these articles include pictorial diagrams that give you a view of the wired circuit right down to the last solder joint. Most often, these projects are either gadgets (unabashedly so) that are fun to build and own, or useful electronic devices that can't be purchased ready-made (or are much less expensive than commercially made equivalents).

A photoelectric "eye" to guard a house, a timer that measures the velocity of a bullet, a repeating photographic strobe light, a "computer" that plays a game, a xenon timing light to tune car engines, a circuit that flashes Christmas tree lights in time to music, a loud hailer/intercom for use on a small boat—these are all projects that I have designed. Which are gadgets and which are useful? You answer that one!

When you tackle a project from scratch, you'll find yourself engaged in a wide range of activities, everything from metalworking (when you cut the necessary holes in a chassis or cabinet) to mechanical assembly (when you mount the parts) to soldering to troubleshooting (only you can work the bugs out). Project articles usually include enough how-to instruction to smooth over the tricky steps, but a successful project builder ought to have a dash of ingenuity.

Want to assemble ready-to-wire kits? You've got lots of company: thousands of people who, each year, build color TV sets, electronic organs, and super stereo amplifiers—step by step by step.

There are two main reasons to build kits: money-saving and fun. The latter needs no explanation, so let's talk about money. Kits cost less than identical factory-built equipment in the short run because you eliminate a share of the labor costs by putting them together yourself. In the long run, you save even more, because you are able to service a kit-built device yourself, eliminating most expensive repair bills.

Yes, you can assemble a kit—even a complicated one—and make it work. I hope that statement should remove the doubt that has probably been gnawing away at your resolve to buy a kit. Of course, most builders find it's best to start with a simple kit—just to get your feet wet. But, if you've been dreaming about a color TV kit, go ahead and order it. It's surprising but true that assembling most familiar electronic equipment doesn't require great manual skill. Tightening nuts and bolts, stripping wires, soldering, observing color codes can easily be learned by anyone who can turn the pages of a book. What book? The excellent instruction manuals supplied with every kit produced by a reputable kit manufacturer.

When you build a kit, you won't learn much about the electronics behind the device's operation (unless you read the theory section included in the back of most manuals), but you will learn a great deal about handling electronic components. That's why I consider wiring a kit to be an excellent introduction to working with electronics, even if you plan to specialize in from-scratch projects. The selection of kits to choose from is enormous, encompassing test equipment, stereo and hi-fi gear, shortwave, Citizens Band and amateur radio sets, television sets, musical instruments, marine electronic gear, and even electronic computers.

Projects? Or kits? If you haven't made up your mind, why not try both? You'll find pertinent information throughout this book that will help you get started.

A Few Words About Safety

It's no fun to read about safety, but it's even less fun to be shocked or burned, or struck in the eye by a metal filing. All of these are potential hazards when you work with electronics. So, for safety's sake, keep the following suggestions in mind when you work:

• *Power-line voltage can kill. So can other higher DC and AC voltages found inside many live electronic chassis. Don't work on any exposed AC-powered chassis unless you are sure that the power cord is not connected to an AC receptacle.*

• *When troubleshooting procedures require that an AC-powered circuit be operating, exercise extreme caution. Make sure that no metal tools lying on your workbench can come in contact with any part of the chassis, and be certain that the chassis is resting solidly on the bench—in no danger of tipping over—before you plug it in. Also, make sure that you are firmly seated, so that an accidental shock won't cause you to fall, and possibly injure yourself.*

• *If possible, obey the old lineman's rule: keep one hand in your pocket when the other is near a live electrical terminal.*

• *Bear in mind that large-value capacitors found in many circuits store a significant quantity of electrical energy. And, depending upon circuit design, they may be able to deliver a painful shock a long time after the power has been turned off.*

• *Never reach blindly for a soldering iron. You may pick it up by the tip!*

• *Always wear workshop safety glasses when you work with power or metalworking tools.*

• *Note that diagonal wire cutters will propel small wire clippings with considerable force, over a considerable distance. Always direct the depressed side of the cutters away from your—or an observer's—face.*

Building From a Kit

It's only stretching the history of electronics very slightly to say that the very first significant consumer electronics product, the crystal set AM radio, was really a kit. Ask anyone who owned a crystal set, and he'll tell you that tuning in a station was an elaborate operation. Actually, although many crystal sets were sold factory-assembled, countless owners built their own by hooking together an antenna coil, a tuning capacitor, a crystal (complete with "cat's whisker contact"), a filter capacitor, and a pair of earphones to create a receiver circuit similar to the modern version.

During the fifty-plus years of our electronic age, kit development has kept pace with the developing electronic technology. This isn't surprising when you consider that the very nature of electronics construction makes ready-to-wire kits possible, practical, and extraordinarily popular. Most familiar electronic products consist of complex components connected together in a relatively simple fashion. Thus, kit manufacturers can gather together the necessary components to build almost any electronic device, toss in a set of instructions that explain what gets connected where, and let the purchaser put the device together by himself, at home. This, in a nutshell, is the idea behind the unique phenomenon of ready-to-wire electronic kits.

And "unique phenomenon" is about the only way to describe the huge variety of kits that are available to the man who works with electronics.

Audio equipment. You can assemble a simple audio amplifier or a super-fidelity stereo receiver (AM/FM stereo tuner plus stereo amplifier) that is the performance equal of any factory-built stereo rig. In between, you can choose among an almost endless assortment of amplifiers, loud-

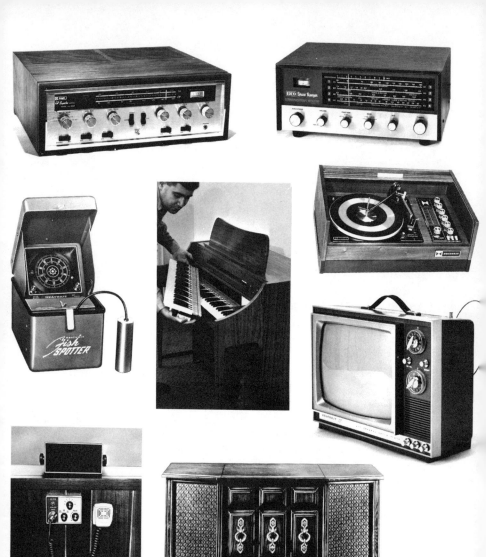

The objects you can build from kits are so numerous it would take an entire book to list them. A few shown here are, clockwise from upper left, radio stereo receivers and amplifiers, shortwave receivers, stereo compacts, television sets including the most advanced color units, console stereo radio and record players, boat communications equipment, fish spotters, electronic organs, and on and on.

speakers, AM and FM tuners, preamplifiers, tape recorders, audio test equipment, and accessories.

Electronic test equipment. Name a piece of equipment— VOM, VTVM, TVM, signal generator, oscilloscope, power supply, component tester, distortion meter, component substitution box, and many others—and chances are, you can build it from a kit.

Communications gear. Whether you're an amateur radio operator (or "Ham"), a Citizens Band radio user, or a short-wave listening buff, the kit catalogs include equipment you'd like to build and own. Transmitters, receivers, station accessories, antennas—all of these are available in kit form.

Electronic musical instruments. Several years ago, kit-built electronic organs were first introduced. Now, the roster of musical instruments in kit form includes electric guitars, electronic rhythm generators, high-power instrument amplifiers and loudspeakers, electronic reverberation and tremolo circuitry, and electronic percussion generators.

Marine electronics equipment. You can navigate your boat with the help of a kit-built direction finder; avoid shallow water (and find fish) with a kit-built electronic depth sounder; chat with nearby boats via a kit-built loud hailer/intercom; and safeguard your boat with accessories such as electrolysis detectors and fuel-vapor alarms.

Home entertainment products. The list includes television sets (both black-and-white and color), portable radios, and clock radios.

Automotive electronics gear. You can build solid-state ignition systems, tachometers, engine analyzers, auto burglar alarms, tune-up instruments (such as timing lights and dwell meters), and mobile communications equipment.

Photographic equipment. What well-equipped darkroom can be without an accurate enlarger timer, or an enlarger exposure meter, or a CdS exposure meter?

Electronic gear for your home. Typical kits include room-to-room intercom systems, garage door openers, home surveillance systems, power-tool speed controls for your workshop, and even high-intensity lamps.

Miscellaneous equipment. There's an inverter that lets you power AC appliances from a 12-volt battery (it's for camping trips); an analog computer, if you are mathematically inclined; a photoelectric relay to watch over a doorway; a remote-control rig for model airplanes; and, of course, an old faithful crystal set (some things never grow old!).

What Is a Kit?

It's really a lot more than simply an unassembled electronic device. Sure, it's a collection of electronic components, a chassis, possibly a few printed circuit boards, maybe a wiring harness, hardware, wire, assorted odds and ends, and a detailed instruction manual, but this brief list leaves out a vital point: A ready-to-wire kit is designed to be assembled by inexperienced people. This means that kit chassis layouts are designed with an eye to avoiding tight corners crammed full of components; that the assembly sequence (the order of assembly steps) is designed to insure minimum confusion; that liberal use is made of color-coded wires to reduce the chance of wiring errors; and that circuitry is designed so that it can be adjusted and/or aligned without the use of external test equipment (in the great majority of kits sold today).

This last point is especially important when you build any kit that incorporates radio-frequency circuitry; in other words, any type of radio receiver or transmitter, or TV set. The alignment of the tuned circuits in r.f. stages is a critical factor in overall performance. In fact, r.f. circuitry will not work at all if sufficiently misaligned, and even slight misadjustments will severely affect performance.

One of the reasons that kit building is a popular pastime today is that kit manufacturers were able to overcome this handicap. Usually, critical r.f. adjustable components (such as r.f. coils) are prealigned at the factory. The instruction manual explains how to touch up the adjustments for peak performance when the device is completed. Often, kit manufacturers supply critical r.f. stages (such as TV and FM tuner stages) completely prewired and aligned. The builder treats the stage, which is usually a sealed chassis or a small printed

This kit by Heath, an AD-27 compact sound system, allows the builder to test its performance after the chassis is wired and before final assembly. Here a metal probe is inserted into the right auxiliary input socket to produce a "hum" from the right speaker system.

Kits are usually constructed for maximum ease of testing and troubleshooting. Here the tuning section is pivoted upward to check components and wiring underneath.

circuit, as a large-component "black box," mounting it and wiring it to the other components without worrying about critical adjustments.

The instruction manuals that accompany most kits contain elaborate step-by-step instructions that are concise, easy to follow, and logical; pictorial diagrams are normally included to clarify the written instructions, and to illustrate unusual assembly techniques.

Most manuals also include sections that describe in detail how the particular circuits operate, and also, present troubleshooting hints if the circuit ever requires service.

Be warned, though, that the manuals are rarely perfect. Minor errors (often typographical) are common, although errors serious enough to cause circuit miswiring are normally spotted early in the life of any kit, and are corrected by errata sheets you'll find tucked into the manuals. And we've found that many manuals leave obvious things unsaid. We'll say a lot of them in the following chapters.

Manufacturers of kits often supply critical r-f stages completely prewired and prealigned. Instructions are given for final touch-up adjustments for peak performance. Here a plastic alignment tool, supplied with the kit, is used to adjust tuner with "flag" of tape to tell the builder how far he has turned the tool.

How Good Is Kit-Built Electronic Equipment?

It can be as good as the best equivalent commercially built gear you can buy, if you take the time and effort to assemble it according to the instructions. By "equivalent" we don't mean "same price"; we mean circuitry that is designed to the same performance standards. In a few product areas—most notably color TV and stereo high-fidelity—kit-built devices are up to state-of-the-art standards. In short, you'll have difficulty finding commercially built gear that can top the performance of these kit-built circuits.

But, keep in mind two potential shortcomings of kit-built gear before you buy:

1. In many product categories, kit-built equipment may not have as high a resale value as commercially built gear, simply

because a prospective purchaser of used equipment is in no position to judge the craftsmanship of an electronic hobbyist. Thus, building a kit makes more sense if you intend to keep the equipment for a long time, rather than plan to trade it in for higher-grade gear in a short time.

2. It is often difficult to find professional servicemen who will service a kit you have built. Many pro technicians fear that circuit troubles are probably caused by hard-to-troubleshoot wiring errors. Thus, plan to service your kit-built device yourself, or be prepared to return it to the kit manufacturer in case of trouble.

Happily, servicing rarely becomes a difficult problem because kit-built devices are usually easy to service. Most kit builders discover that assembling the circuit gives them enough familiarity with the various components to follow the troubleshooting hints outlined in the manual. And, where circuitry requires periodic adjustment to maintain peak performance, kit manufacturers often build test-equipment circuitry into the circuit to make this adjustment easy, or else they devise alignment procedures that do not require test equipment to perform.

Incidentally, the troubleshooting techniques explained in Chapter 10 will help locate any circuit flaw that is too elusive for the troubleshooting hints given in a kit instruction manual.

Will My Kit Work When It's Finished?

This is the question every kit builder asks himself when he starts working (and every potential builder asks before he orders a kit). The best answer I can give—based on the personal experience of wiring over 150 different kits—is *probably*. Here's why:

71 percent of the kits I built worked perfectly from the moment they were turned on.

22 percent didn't work because of one or more minor errors I had made. In most cases, a brief period of basic troubleshooting spotted the error(s).

4 percent didn't work because of bad components. Here

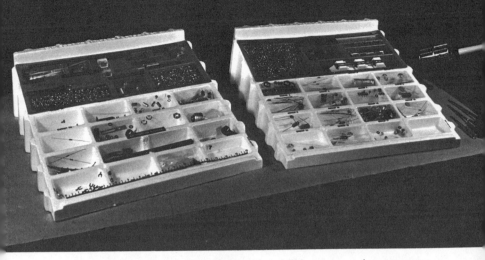

Parts for this kit, the Scott LR88 AM and FM stereo receiver, come packed in plastic trays. Small bins in the trays contain components, preassorted and ready for use in each step as called for in the instruction manual.

again, following the manual's troubleshooting hints usually pinpointed the faulty part.

3 percent didn't work because of an error later traced to an incorrect instruction in the manual (all of these were in brand-new kits).

As you might expect, nearly all of my minor errors were careless errors which could have been avoided. However, I have observed that many novice kit builders make so-called minor errors, not so much out of carelessness as out of confusion caused primarily by lack of experience:

- It's very easy to confuse color-code stripes on resistors.
- It's very easy to reverse terminal wiring to a tube or transistor socket.
- It's very easy to "forget" to solder a component lead to a printed circuit board, or to not notice a short-circuiting solder bridge lying across two adjacent conductors.
- It's very easy to overlook a cold-soldered joint hidden among a series of good joints on a crowded terminal strip or tube socket base.
- It's very easy to mix up the wires bundled together into a wiring harness because of the similarities of their color-coded insulation.

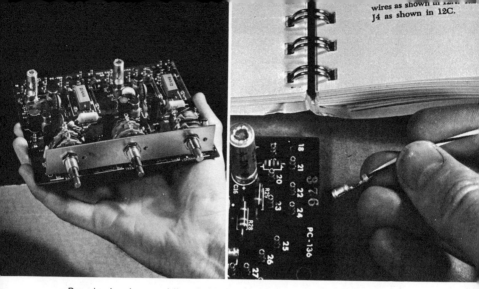

Prewired subassemblies (left) for the Scott LR88 save time plus painstaking work and are ready to be wired into the kit. The tuner front end also is prewired and factory aligned. Solderless connectors (right) are helpful in the final stages of wiring the assemblies. Tiny metal collars are fitted onto the ends of the wires and are slipped over proper posts on the circuit board and snapped into place.

● It's very easy to damage certain heat-sensitive components with poor soldering technique.

The point of all of this is not to discourage you from attempting to build from a kit, but to demonstrate that in all probability you will make a careless error when you wire a kit. Consider it as a payment of your fine for disobeying the law of averages by not keeping a closer watch on your actions as you work. But don't consider your kit "finished" until the trouble spot is located. Errors and faulty components are a natural and normal part of any assembly process. Even professional technicians make mistakes; that's why commercial electronic factories have quality-control inspectors to reexamine and repair finished chassis that don't work.

Above all, don't place your ego on the line. More than one first-time kit builder has hidden a nearly finished kit on a back shelf of a storage closet just because it didn't work on the first flip of the power switch. It's a shame, since a bit of troubleshooting could turn a black sheep into a prized conversation piece.

Judging a Kit Before You Buy

The kind of judging we're talking about here doesn't have anything to do with evaluating the electronic performance of a finished kit. In this regard, you must treat a kit-built item in the same way as a factory-built device, and consider the published specifications, the manufacturer's reputation, various test reports published in magazines, and (with some kits) actual demonstrations in local retail stores.

In this section we'll be looking at those characteristics that can help you judge whether or not a specific kit merits your consideration—based upon your specific requirements and/or preferences.

Practically speaking, there are four characteristics possessed by every kit that can—when considered together—let you judge:

1. Money you will save by building the kit, compared to an equivalent factory-built device.

2. Length of time it will take you to assemble the kit, based on your level of experience in working with electronics.

3. Complexity of the assembly stages, in comparison to the construction techniques you are familiar with. To put it another way, how many new skills will you be asked to master in order to complete the kit?

4. What is the risk of making at least one error that will require returning the kit to the manufacturer for troubleshooting? Or what are the odds you'll make a mistake you won't be able to correct yourself?

Clearly, all four factors are closely interrelated. A kit that promises to save a lot of money probably is complex to assemble. (Your savings represent the labor costs that would be expended in factory-wiring the device, less the costs of preparing the manual and kit packaging.) This complexity may be demonstrated in long hours of repetitive, straightforward wiring (as in assembling an electronic organ), or in a number of relatively difficult assembly steps, which are time-consuming because they require great care (as in assembling a sophisticated stereo receiver). The risk of your making a serious error is different in both situations.

How to Wire a Kit

A good part of what we say here can be found in the generally excellent instruction manuals issued by the major electronic kit manufacturers—if you read between the lines. We
don't claim the manufacturers are trying to keep any helpful
information from kit builders. But the hard fact is, that many
of the mistakes that first-time kit builders make can be traced
to the things that the manuals don't come right out and say.

For example, the elaborate pictorial diagrams are not to be
considered as totally accurate except in the way they show all
of the various connections within the chassis.

How so? For one thing, the artist can subtly alter the size
of a terminal lug so that a half-dozen drawn-in-ink wires can
each be looped around the terminal with room to spare.
Doing the same trick with six real copper wires and a real
terminal lug, while avoiding short-circuits against adjacent
lugs, is a real challenge to an experienced kit builder. The
problem is creating a mechanically and electrically secure
joint that includes all six wires, without spilling over against
adjacent lugs. We'll say more about this shortly.

Also, the diagram may show all interconnecting wires
neatly bent in sharp turns, and dressed cleanly against the
chassis. As you might expect, ink lines are easier to handle
than copper wires, so that a really neat wiring job will still
look sloppy in comparison to the diagrams, a fact of life that
shoots down the morale of many an uninformed beginner.
Worse yet, if a beginner tries to achieve the snappy-crease
look by making sharp bends in the wire, he may accidentally
break the conductor, and create one of the hardest-to-find circuit flaws.

Throughout this chapter, we will discuss a variety of unsaid
items like the above. It has been our experience that each has
been responsible for a builder's goof.

What to do when the kit arrives. The point to keep in
mind is that kit packaging is designed with as much ingenuity
as kit circuitry. Once you start unpacking—according to the
schedule called for in the instruction manual—you'll be
amazed at the volume of parts and waste paper that comes

out of the box. Thus, before you begin, have an assortment of several deep trays at hand (baking tins and muffin pans are fine) to hold small components and hardware. Also, place all discarded wrappings in a clean cardboard box, but do not throw the box away until the kit is finished. It often happens that a nondescript hunk of brown paper turns out to be a critical template or equally important part. Follow these steps in unpacking the kit:

1. Carefully open the carton and remove nothing but the instruction manual.

2. Locate all *errata* sheets (if any) and make all indicated changes in the manual.

3. Gather together the various documents, warrantee cards,

---

1 When you open the carton containing your kit, remove the instruction manual and become familiar with the introductory material and general instructions on tools, wiring, soldering, and assembly procedures before beginning to remove the contents.

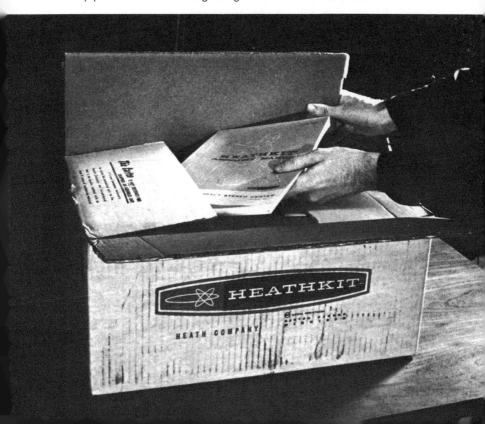

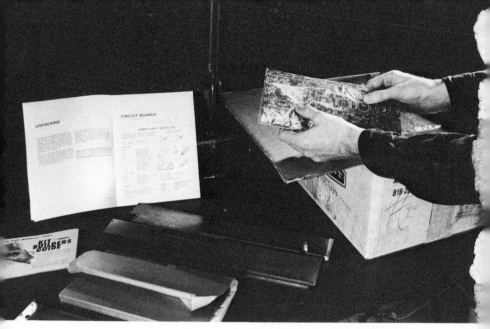

2 Remove each item carefully and place all parts on a table. Smaller components usually come packaged in smaller containers, which should not be opened until the manual, in step-by-step instructions, tells you to do so.

---

3 After opening the package of the first assembly stage, examine each component and check it off in the parts list. Muffin tins, TV dinner trays, or other compartmented containers keep the components in good order.

4 Study resistors carefully to become familiar with the color coding on them, and be very sure you have the correct color combination (grey-red-brown, for example) when called for in the instruction manual.

5 In this kit, construction begins by soldering resistors and capacitors to a power-supply circuit board, each step given in instructions. Wire leads of each component are bent at right angles and placed through holes in the circuit board.

6 The circuit board is turned over and wire leads are bent slightly so the component will stay in position during the soldering step.

7 Each lead is soldered to the circuit board (for complete instructions on soldering, see Chapter 6).

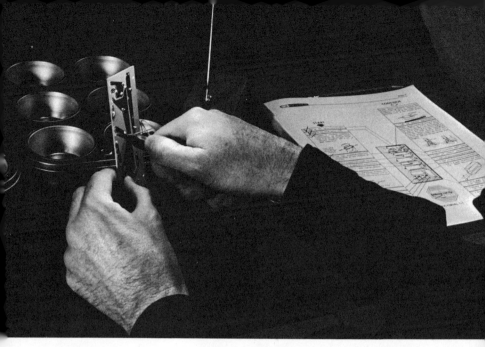

8 After solder has hardened, excess lead lengths are cut with diagonal cutters. Turn the board partly away from you as the leads tend to fly off at high velocity.

---

9 Some steps require one component to be soldered to another, requiring a "third hand." Locking pliers used as a vise fill the bill.

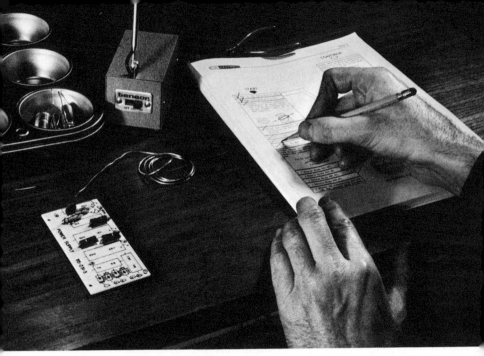

10 After each step is completed, it is checked off in the manual.

11 When finished, examine with considerable feeling of satisfaction your first completed circuit board.

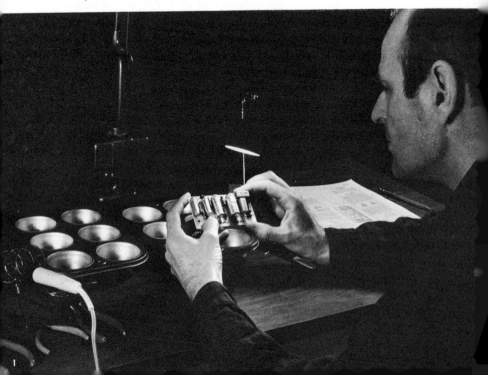

and other items that fell out of the instruction manual, and put them away in a safe place.

4. Unless you intend to start building there and then, don't unpack the carton. *Note*: If your kit arrives by parcel post or express, its carton will carry a printed legend urging you to "examine the contents immediately for possible damage." Unless the carton is visibly damaged, leave well enough alone—and the carton unpacked. Chances are you'll find a recommendation printed early in the manual telling you to "leave all components packed in their numbered containers until the instructions tell you to open them."

Tools you'll need. Almost paradoxically, it usually requires a broader selection of hand tools to assemble a ready-to-wire kit, than to build a project from scratch. This is because kit manufacturers make use of a much wider variety of hardware than a home builder would, and they often mix several kinds of construction techniques in a single kit—for example, part printed circuitry, part terminal lug construction, and part point-to-point wiring is a common mix.

Although you can build a kit with a minimum selection of tools, overall appearance usually suffers. Screw-head slots are gouged by screwdrivers of the wrong width, for example.

Add the following items to the basic tool kit outlined in Chapter 3, and you'll have a kit-builders toolbox:

$3/16''$ cabinet-blade screwdriver; $3/8''$ standard blade screwdriver.

Miniature long-nose pliers.

Miniature diagonal cutters.

Hemostat (or locking forceps, as described in Chapter 3).

$1/4''$, $3/16''$, and $11/32''$ nutdrivers.

Spring-type pickup tool.

Setting up a working schedule. Building a kit during short working periods—say an hour each evening—takes more time than building the same kit in one or two long, continuous work sessions, because you necessarily waste time starting and cleaning up during each short period. However, in my experience, working continuously for long periods increases the possibility of making careless errors.

The solution is a long work session of four to five hours,

To keep parts in good order, Heath suggests cutting a corrugated box at an angle and inserting leads of resistors and capacitors into the edge to hold them. Be sure, however, that small components will not fall into the corrugated slots. If so, you'll have to dig them out at the bottom.

---

consisting of several short work periods separated by shorter breaks (short enough so that no start-up time is needed to get back to work and you don't even have to turn the soldering iron off). Here are the relative work and rest periods I have found most efficient for different kinds of activities:

Type of Work	Work Period	Rest Period
Printed circuit assembly	30	10
Wiring harness hook-up	30	10
Point-to-point wiring	20	10
Mechanical assembly	45	15

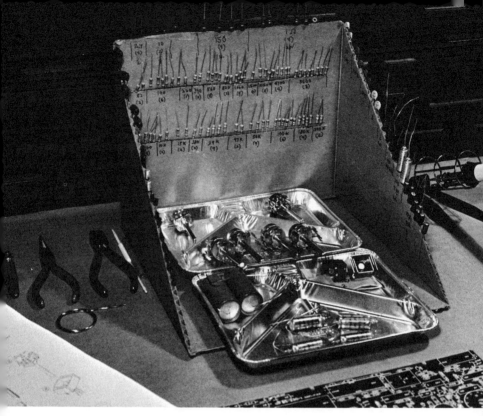

A better storage method is to cut slits along the back of the box to store small resistors so they won't slide down. Some beginners prefer to label resistor values on the box to make identification easier later and as a double check.

Another way to prevent small components from falling into corrugated slots is to place a piece of tape over the edge. Insert the lead lengths of components through tape and into the slots.

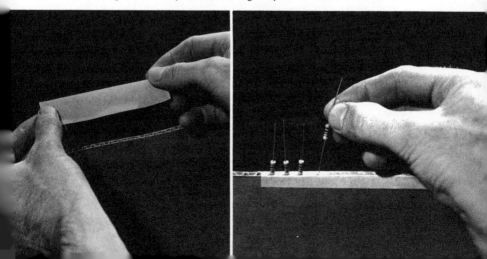

Common errors in mechanical assembly. The following errors are frequently made by kit builders. Avoiding them will save you a lot of time and work later.

• *Mixing up hardware.* Too-long screws on crowded chassis may cause short-circuits; too-small bolts on large, heavy components may shear off.

• *Forgetting lock washers.* Vibration may, in time, loosen machine screws, allowing chassis-mounted components to move, and possibly short-circuit.

• *Using excessive torque to tighten screws and mounting nuts.* May crack fragile phenolic and plastic parts; distort switch and control bodies; deform power semiconductor cases, with subsequent reduction of heat-sink contact that might lead to device failure; fracture insulator wafers mounted beneath certain audio jack assemblies; strip screw threads tapped in chassis.

• *Failure to label all chassis-mounted components* (as recommended in most manuals). May result in misplacement of mirror-image terminal-strip leads in point-to-point wiring.

When removing or attaching a speaker, hold the screwdriver carefully with both hands to be sure the top doesn't slip and damage the delicate speaker cone. Better yet, use a nutdriver (right), if possible, as it can't slip as easily.

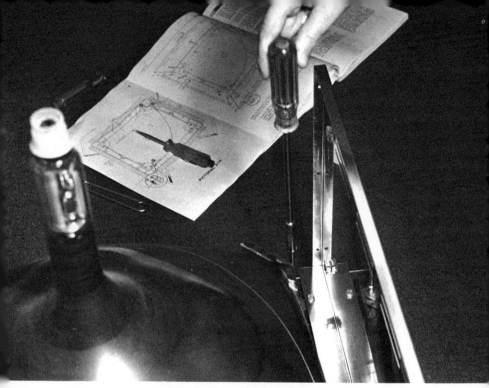

An extension screwdriver or nutdriver can be helpful when the instructions call for screw tightening deep in hard-to-reach corners of a chassis, and can help protect vulnerable components (such as the neck of a color picture tube) from accidental damage. If you must use ordinary-length tools in situations like this, remember that a carelessly swung screwdriver can scratch decorative surfaces, and wreck fragile components.

Terminal identification errors lead to serious wiring errors, possibly to component damage.

Pinpointing overloaded terminal lugs. It's possible only if your kit's manual lists the number of wires to be soldered next to each solder instruction (such as S2 or S5). Scan through the instructions before you begin to solder, and note the locations of any terminals (terminal lugs, switch terminals, tube socket lugs, control terminals, etc.) that will carry four or more wires. Mark an X next to the terminal on the chassis. This mark should remind you to be especially careful as you add wires to the terminal.

If possible (without violating the kit designer's lead placement plan) route incoming wires so that all come into the terminal in parallel paths.

Keep the length of exposed copper at the end of each incoming wire as short as possible—preferably less than ¼ inch.

If the terminal will hold five or more wires, do not secure each mechanically to the terminal. Pass the end through, and give it a 90-degree bend to prevent the wire slipping back out. Once the terminal is soldered, cut off excess wire close to the terminal. Watch for possible short-circuits.

To solder or not to solder. The instructions that tell you when to solder and when not to solder have two goals in mind: First, an unsoldered terminal looks unfinished, hence you know you will have to add additional wires before the chassis wiring is complete. And, second, the kit designers are afraid that clumsy soldering technique would fill a terminal with solder, preventing the placement of additional wires, if the instructions directed builders to solder each wire in place in turn.

If you need the psychological crutch of an unsoldered terminal to keep you from making mistakes—okay. Follow the instructions to the letter. The same thing goes if you can't master the technique of applying a little bit of solder at a time to a terminal.

But, once you are beyond this stage, consider the following:

1. Soldering each lead in turn yields a neater wiring job since you don't have to provide a secure mechanical bond to hold the lead in place until you get around to soldering the terminal.

2. Soldering each lead in turn guarantees that each is part of the final multi-connection joint. When many wires are soldered in place at one time, the ones at the bottom of the heap may not end up soldered to the terminal.

Soldering to ground terminals (or chassis lugs bolted to the chassis). The only way to do the job properly is to deliver enough heat to the lug so that it, and the wires passing through it, are heated to well above the melting point of solder. Obviously, since the chassis is a large heat sink capable of absorbing large amounts of heat, a high-wattage iron or

Be sure to use a high-wattage soldering tool—either an iron or gun—when instructions call for soldering to the chassis. Usually, these soldered connections are establishing ground connections which are critical to proper circuit operation. Low-wattage irons suitable for circuit board soldering simply can't deliver enough heat fast enough to produce a good joint.

gun must be used. Don't attempt to use the same low-wattage iron that does a perfect job on printed circuit boards and terminal strip lugs. It will—at best—produce a cold-soldered joint. Since chassis ground connections are usually critical, this is bound to ruin device performance.

Wiring-harness worries. Here again, the actual wiring harness supplied with your kit will not look like the trim, smooth, perfectly dimensioned harness depicted in the diagrams. Don't waste time trying to iron out the wrinkles, and above all don't yank on the various "breakouts" to make their relative locations duplicate the diagrammed harness.

Coaxial cable in miniature. If you aren't careful, miniature coaxial can be the ruin of any kit it's used in, and leave you with a rugged troubleshooting problem. The hazard is overheating during soldering. The small diameter ring of insulation can melt if excess heat is delivered to either the outer

Protect small-diameter coaxial cable by fixing a heat sink to the shield before you solder it in place. A common hair clip (steal one from your wife) works fine. It draws off excess heat before it reaches, and melts, the cable's inner insulation (which would short-circuit the cable).

shield conductor (most usual cause) or the inner stranded core. Consequently, the conductors short-circuit. Most effective cure is to heat-sink the shield close to the cable body as you solder it in place. A hemostat makes an ideal heat sink; or you can use a commercial semiconductor-lead heat sink. Normally, if you work quickly with a small-tipped, medium-wattage soldering iron, you can solder the central conductor in place before the insulation under the shield begins to melt.

Make it easy for yourself. Whenever you mount value-labeled components (not color-coded) on a printed circuit board or rotary switch, or between terminal strips, position the part so its label is facing upwards or outwards. You'll save a lot of looking around corners in case troubleshooting is required.

When to call in an assistant? Whenever an extra pair of hands will save you time. A few examples:

When mounting heavy components on a large chassis. Your assistant positions the part against the chassis, while you insert the mounting bolts.

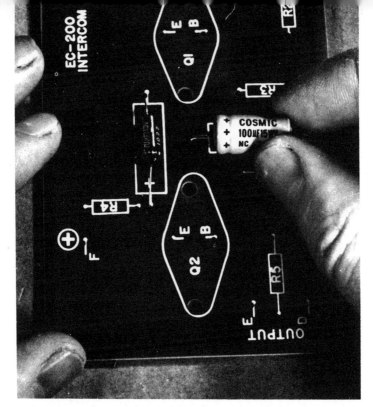

When placing value-labeled components on circuit board, be sure to place them so labels are facing upwards for easy reading in case checking is necessary later.

If the job is a big one, like unpacking a color TV picture tube, and placing it into position on its ''mask,'' be sure to get a helper before you begin.

When mounting light components on a large chassis. Your assistant eliminates the need to twist your arm into a pretzel while you simultaneously position the part and fiddle with nuts and bolts on both sides of the chassis.

When doing almost anything to a large chassis, including soldering, balancing, rotating, lifting, and moving.

Working with printed circuits. Unless you are experienced with PCs, don't follow the manual's recommendation to solder leads after you've inserted several components. Solder as you mount each one. Soldering in a forest of leads is confusing, and may cause solder "bridges" (solder blobs short-circuiting adjacent conductors on the foil). Beginners often miss soldering one or more leads among the many; awkward board handling may let some components slip away from the board before their leads are soldered (thus leaving a gap between the component body and the PC board), upsetting total PC board height, and possibly contributing to short circuits when the board is mounted in a tight chassis layout.

Lingering with a soldering iron at one point on the copper foil will overheat the foil and may cause it to peel away from the insulating base. When you solder, work quickly with a hot, well-tinned iron.

Don't use a large-tipped soldering iron. The wide point may contact several solder "pads" (wide area around component hole) simultaneously, melting solder and developing a solder bridge with nearby conductor strips. Be sure to use a narrow iron point (approximately ⅛ inch is ideal for most PC board work).

Avoid flexing and/or bending the board. Copper foil isn't as stretchable as the board is bendable. Flexing may rip breaks in the foil.

How to spot "stupid" mistakes. Any of the following clues should alert you to start looking for an error:

You "run out" of a component of specific value called for in the next instruction. Chances are that you installed it by accident a few steps earlier.

You have a component "left over" after completing all steps. Probably, you omitted a step awhile back. To locate

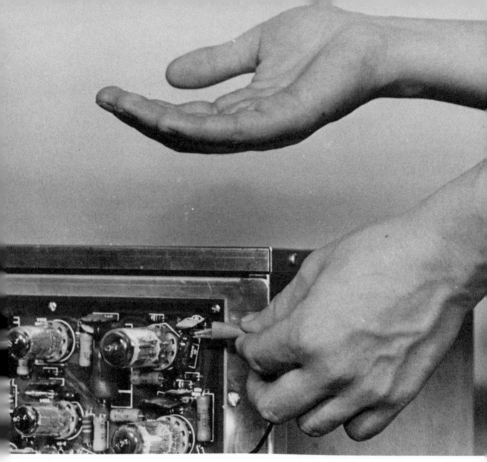

When you must work with exposed high-voltage terminals, use an old power lineman's trick: keep one hand cupped upwards away from the chassis (so dangling fingers don't complete an accidental circuit that could give you a bad shock), or keep your free hand in your pocket.

where the part belongs, dig out the master parts list, find the number (or numbers) of all part(s) of the "extra's" value, and mark each on the schematic diagram. Then, compare the chassis with the diagram to pinpoint the vacancy.

You have an excessively long excess lead length to deal with after you connect the free end of a wire that is already soldered to another point in the chassis. It's probable you've connected the free end to the wrong point, since lead length specifications instructions rarely leave more than ½ inch of

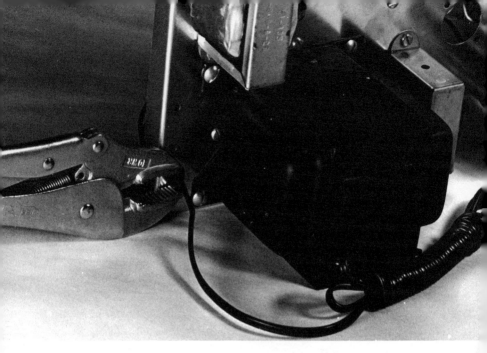

An easy way to install a power cord "strain relief" is to use a pair of locking pliers to compress the gadget around the line cord. Once compressed, it will snap into the chassis mounting hole with a little pushing.

excess length (measured after you have routed the wire along the path shown in the pictorial diagrams).

When you find an "error" in the instructions, don't believe your eyes until you've cross-checked the written instructions with the pictorial *and* with the schematic diagrams. Serious manual errors are rare; an error on your part is more likely.

How to install AC line-cord "strain reliefs." Before you try (vainly) to force the relief through the hole in the chassis, form the line cord into the relief by squeezing the two halves of the relief with a pair of pliers ("hose clamp" pliers, an automotive accessory, or lockjaw pliers, are ideal).

Mounting knobs on round shafts. If control shafts that are fitted with setscrew fastening knobs don't have "flats" to insure good grip between knob and shaft, provide them. Use a narrow Swiss Pattern file to form the flat. Here's how to locate the exact spot to file: After the kit is finished, tighten the setscrews of the knobs in question as tightly as you can without cracking the knob. Remove the knobs and examine the control shafts for roughing caused by the setscrews. Carefully file away the roughing, always keeping the file perpendicular to the roughing's direction.

If the kit doesn't work, don't panic. If everything looks like it's in the right place on the chassis, the chances are that one of the following gremlins is at work (listed in order of decreasing probability):

- One or more solder joints are bad or incomplete.
- One or more leads aren't soldered in place on a PC board.
- One or more wires or component leads aren't connected to the right terminals.
- Two or more components are swapped.
- You didn't complete (or follow) the adjustment, alignment, or tune-up steps.

Don't be surprised if your kit works perfectly the first time you turn it on—but don't necessarily expect it to. Even if you've followed all instructions perfectly, and all components are working, there may be a series of critical adjustments (such as conversion adjustments on a color TV) that must be performed before circuit performance is "normal."

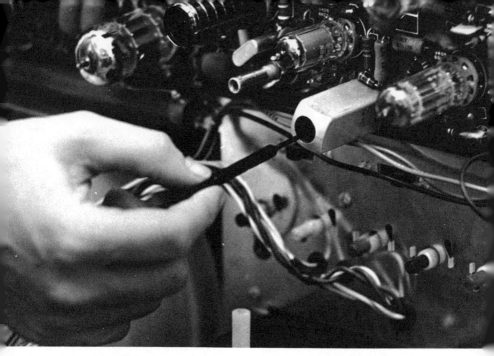

The whole idea of an insulated alignment tool is to keep your hands (and the ''body capacitance'' they represent) away from critical adjustable components such as r.f. coils. Don't defeat the purpose by ''choking up'' on the tool; grip the tool lightly as far from the component as possible.

If you must remove a tube, prevent tube pin damage by gripping the envelope firmly (wait for a hot tube to cool first), and pulling straight backwards out of the socket.

● One or more parts is faulty.

If you locate an error on a PC board, be prepared to sacrifice the components involved to protect the board. *Note:* Repeated soldering and unsoldering to the foil on a PC board is likely to peel the foil away from the backing. Without proper de-soldering tools it is difficult to remove a multi-lead component (two or more leads) from a PC board. This is because a conventional iron only lets you melt the solder surrounding one lead at a time. It's almost impossible to move the iron quickly enough to keep all leads melted so you can withdraw the part. And the board will likely be damaged while you are trying, and the component may be heat-damaged anyway.

Thus, if you've accidentally swapped two easy-to-get components (resistors, capacitors, transistors), or made a lead polarity error with an electrolytic capacitor, transistor, or diode, cut off the leads at the component body, and remove each from the board individually. Obtain replacement parts locally, or from the kit manufacturer, and start fresh.

Kit Manufacturers

Heath Company
Benton Harbor, Michigan 49022
> Wide variety of all types of electronic kits, including TV, audio systems, ham radio, musical instruments, and test equipment.

Allied Radio Corporation
100 N. Western Avenue
Chicago, Illinois 60680
> Comprehensive selection of Knight Kits, as well as a broad selection of kits from other manufacturers.

Eico Inc.
283 Malta Street
Brooklyn, New York 11207
> Test equipment, audio systems, ham radio equipment.

H.H. Scott, Inc.
115 Powdermill Road
Maynard, Massachusetts
  Audio equipment.

Dynaco (Dyna Company)
3060 Jefferson Street
Philadelphia, Pennsylvania 19121
  Audio equipment.

Schober Organ Company
43 West 61 Street
New York, New York 10023
  Electronic organs and accessories.

Delta Products, Inc.
Grand Junction, Colorado 81501
  Automotive ignition systems, burglar alarm devices.

In addition, many large mail-order electronic supply houses
(such as Allied Radio Corporation) carry kits made by R.C.A.,
Raytheon, and Norelco.

Building Your Own Project

The true starting point of any electronic project is a *schematic diagram*, a kind of circuit "road map" that shows how all of the circuit's components are to be interconnected. Chances are, though, that you will have much more information when you assemble an electronic device, especially if you are just beginning to work with electronics. When you follow the directions in a do-it-yourself project article published in a hobbyist magazine, for example, you'll often have *pictorial diagrams* (actual sketches of the assembled circuit), or photographs, to guide you.

Nevertheless, it is important to know how to "start from scratch"—in other words, from a schematic diagram—because this is often the only way you can start. Every year, thousands of clever, buildable circuits are designed, and their schematic diagrams published, in magazines, manufacturer's catalogs, application manuals, and project textbooks.

More important, understanding design principles gives you a valuable edge even when additional information is available. You then have the ability to modify the design of a published project, if you choose, to make the finished project meet your requirements more closely. And you won't be stymied by a chance error in a pictorial diagram (a too-common occurrence) because you can work directly from the schematic.

The design of all electronic projects follows a simple path with three stops along the way:

1. The schematic diagram (and parts list) is analyzed and interpreted to determine the assembly "ground rules" for the particular circuit.

2. The decision about the type of construction and/or wiring technique to be used is made by considering the nature of the circuit.

3. The parts layout and/or chassis layout and/or cabinet layout are planned, keeping the nature of the circuit, and its intended application in mind.

As your circuit wiring experience grows, you will be able to tackle all three steps in an almost simultaneous fashion. But in this chapter we will consider each individually.

Interpreting Schematic Diagrams

A schematic diagram is no more difficult to read than a road map, or the floor plan of a house. If it seems unfathomable to you now, it is just because you are not familiar with the symbols used.

Consider the accompanying diagram of the circuit of a simple photoelectric relay, an "electric eye" device of the type used as door annunciators, or as burglar alarms. Any person or animal breaking the beam of light between the lamp and the photocell makes the circuit activate the relay, and sound the alarm bell. Shown below the schematic diagram is a pictorial sketch of the actual device.

PHOTO RELAY

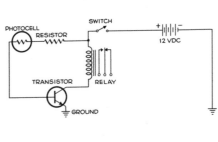

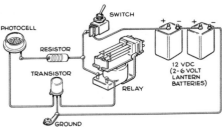

Study the pictorial diagram carefully, for a moment or two, and then think about the problem of conveying all of the information necessary to completely describe the circuit, to completely characterize the maze of components and connecting wires.

The solution is a highly simplified connection diagram that uses easy-to-draw symbols to represent components, and straight lines to represent interconnecting wires, and depicts the scheme of the device. We call this drawing a schematic diagram.

Every resistor in the device is represented by a saw-tooth line, a pictorial representation of the concept of resistance.

Capacitors are indicated by two close but separated lines (each representing a plate of the capacitor), the curved line on the electrolytic capacitor standing for the negative terminal. Some diagrams use + or − symbols instead.

The semiconductor components each have a unique symbol: The diode is a combination of triangle and straight line representing anode and cathode, respectively. The transistor is a circle containing a solid bar representing the structure, and three lines representing the internal elements. (Note that the symbol shown is for a PNP transistor; an NPN transistor symbol has the arrowhead reversed.) The photocell is a composite of a resistor symbol within a circle, indicating that the device acts like a light-controlled resistor.

The relay symbol consists of a coil linked to a switch assembly.

The power transformer symbol is a set of coils coupled together by a core.

By now, you should be getting the idea. The other symbols and their equivalents will be seen in the chart of schematic symbols on page 46.

Interconnecting wires are represented by straight (or occasionally, curved) lines. And here there is a possibility of confusion. It is caused by the rather ambiguous symbols for the joining, or interconnection, of two or more leads or wires, and the symbols for the crossing (but not connecting) of leads or wires.

ELECTRONIC COMPONENTS AND THEIR SYMBOLS

As a rule, and this is a rule that will be followed throughout this book, an interconnection is indicated by a dot at the point two or more wires intersect. A crossing is indicated by a drawn "bridge" of the upper wire over the lower wire, or by just eliminating the intersection dot.

Keep in mind that the whole question of crossing and/or intersecting wires exists only on paper; it's caused by the fact that a schematic diagram is a two-dimensional representation of a three-dimensional circuit. And thus an artist is often required to draw one interconnecting lead—or rather the straight-line symbol for it—over the symbol for another one, even though the actual leads don't interconnect in the circuit.

A schematic diagram by itself is like a road map without a legend; the various symbols don't convey any information about the characteristics of the components making up the circuit. A saw-tooth line could represent a hugh high-power resistor, or a tiny, high-precision resistor. A transistor symbol could stand for a low-power high-frequency unit or a high-power low-frequency device.

This is why a parts list, of some form, must go along with

AMPLIFIER STAGE

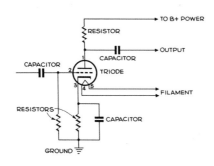

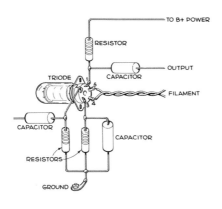

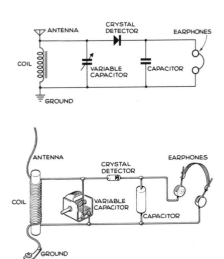

CRYSTAL RECEIVER

a schematic diagram to give it meaning. The two most common kinds of parts list are: (1) an actual list printed next to the diagram; and (2) a set of individual value labels printed next to the appropriate components directly on the diagram.

The ideal situation is to have both; but most often you'll see only the first, with the components keyed to the items in the parts list with brief numerical labels. Thus, the different resistors will be labeled R1, R2, R3, etc; the capacitors C1, C2, etc. These numbers match similar labels next to the individual listings in the parts list.

After you have worked with—and decoded—schematic diagrams and their parts lists for a while, you will find that a glance at a diagram will let you mentally visualize what the actual circuit will look like. In the beginning, though, it helps to draw a crude pictorial diagram as you study the schematic. We've done exactly this with two simple schematic diagrams, the amplifier stage on page 47 and the crystal receiver, above.

Deciding on Type of Construction

At last count there were dozens of different ways to wire and assemble electronic circuits. However, only a few are suitable for home construction:

Point-to-point, on either a 2-dimensional chassis, or within a 3-dimensional cabinet or case. Here, interconnecting wires are routed directly between components with terminals, and small components are supported by their leads when wired to terminal strips bolted to the chassis or case.

Point-to-point on a perforated phenolic board, using push-in terminals as wiring and mounting points. Components can be mounted above and below the board (their leads connected to either the top or bottom ends of the terminals), so we call this a "2½ dimensional" wiring technique.

Terminal-to-terminal on terminal strips. Here, the terminal strips are phenolic (or other plastic) boards equipped with solder terminals mounted close to their edges. You may occasionally find this technique referred to as "military style" construction.

Printed circuits. The interconnecting "wires" are thin layers of copper left bonded on the surface of a plastic board after the excess copper has been etched away chemically. Component leads are run through holes drilled in the board, and

Point-to-point wiring. Interconnecting wires are routed directly between terminals on major components; minor components are supported by their leads when these are soldered in place.

Wiring on a perforated phenolic board. Minor components are secured, by their leads, to push-in terminals that also serve as soldering points for interconnecting wires. Since component leads and/or wires can run across the "top" or "bottom" surface of the board, this technique is also called 2½-dimensional wiring.

connected to the appropriate points by soldering them to the etched copper pattern.

There are kits available that enable you to make your own printed circuit boards at home, relatively simply and inexpensively. However, for most applications, phenolic board construction is easier, faster, and just as effective.

Combination of techniques. The idea here is self-explanatory. The variety of components used in many circuits requires a mix of two or more of the above techniques.

The decision as to which technique to choose when you assemble a project is based primarily on the size, number, and weight of the components, and the complexity of the circuit.

The first step is to gather all of the required circuit components together, in a single group, in front of you.

Next, decide what style and type of cabinet is required, and roughly what you expect the finished device to look like. In some projects, the selection of case material is restricted by the nature of the circuit. (A high-gain amplifier, for example, may have to be built inside a metal case to provide circuit shielding.) This type of requirement is usually stated somewhere on or near the schematic diagram, or possibly in an accompanying paragraph of descriptive text.

I would estimate that 99 percent of all hobbyist-built electronic projects are built into ready-made aluminum or steel boxes, plastic boxes or bakelite instrument cases, or are built on bare metal chassis, without a cabinet.

Home-built cabinets that duplicate many of the store-bought varieties can be made if you have the appropriate skill and shop tools. But, as we say later, it usually doesn't pay to try,

---

Wiring on a terminal board. A rugged wiring technique built around phenolic or plastic boards equipped with metal terminals swaged into their edges. It is suitable primarily for wiring minor components together, since it is difficult to mount major ones on strips.

A commercial circuit breadboard that is basically a large sheet of perforated phenolic circuit board supported in an aluminum frame. The connectors are spring-loaded push-in terminals. Complete breadboarding kits like this are available from electronic supply houses and contain many accessory holders and brackets to mount a wide variety of components.

since the wide assortment of ready-mades lets you build really professional-looking projects inexpensively, and with a minimum of fuss.

Aluminum cabinets are probably most suitable for the vast majority of home-built projects. Generally, plastic boxes and/or bakelite cases can only be used for relatively simple projects, built around lightweight components, designed for applications where ruggedness isn't called for.

Steel is a strong and heavy material, but very difficult to hand work with simple tools. It is most often used to build physically strong and large projects that incorporate heavy components.

Aluminum strikes a happy medium: it is light in weight, relatively strong and stiff, and surprisingly simple to work with. Aluminum cases are available in a wide variety of sizes and styles—probably the widest assortment of any material.

Unfortunately, it's not practical to present a simple set of rules for deciding on case style. You will quickly gain the ability to judge and select after you build several projects. One problem is that the decision is based on aesthetics, the question of suitability, cost, and several other factors. I suggest that you scan the pages of hobby magazines and see how the authors of different electronic project articles have solved the problem for various kinds of circuits.

A good starting point, though (and this is a hint, not a rule) is to use aluminum miniboxes for the first few projects you build from scratch. These boxes are about the most useful enclosures ever developed for home electronic construction. They are available both finished and unfinished, in a great number of sizes and shapes, but all feature the same kind of construction: two formed halves—a bottom and a cover—which interlock to make a closed case.

The next step is to total up the number of major and minor components. A major component is any part or device that is mounted by bolting it to, or through, a surface. For example, the following are major components: power transformers; potentiometers; panel meters; variable capacitors; some power resistors and power transistors; various kinds of sockets and heat sinks. A minor component is any part or device that is mounted by its leads, when they are soldered in place, or inserted into a socket. Examples: small resistors and capacitors; diodes and low-power transistors; small coils.

Finally, compare the component counts with the accompanying construction-technique suitability table. This will give you a good idea of which technique is most practical. As you might expect, the table won't give you a totally correct answer all the time, but it is a good starting point.

You'll note that as yet we haven't said anything about cabinet or chassis size. After you gain circuit-building experience you'll be able to gauge required case size just by studying the schematic diagram and parts list. But, at first, you'll find it difficult to make size estimates before you've actually planned the parts layout. And so, we'll leave this step until later.

Parts Layout and Breadboarding

It is pretty much a truism that a well-laid-out circuit—in any type of construction—is good-looking. It will be clean, neat, and have a well-organized look about it. But a layout designed only for good looks won't necessarily perform well. Many electrical factors must be considered, too. This is why you must avoid the temptation of translating a schematic diagram too literally into a parts layout. The artist who drew the

CHOOSING A CONSTRUCTION TECHNIQUE

CIRCUIT COMPLEXITY	COMPONENT TYPE RATIO	CONSTRUCTION TECHNIQUE	COMMENTS
Simple—less than 15 components	Mostly minor components, resistors, capacitors, two-lead devices	Terminal strip, PC, perforated board	For simple, battery-powered projects
Simple	Mostly minor components with several 3-lead and 4-lead devices	Perforated board, PC	For typical simple solid-state devices
Simple	Equal mix of major and minor components	Point-to-point	Terminal strips as minor soldering points
Simple	Mostly major components	Point-to-point	None
Moderate—less than 30 components	Mostly minor components	Perforated board, PC	For typical moderately complex solid-state projects

Moderate	Equal mix of major and minor components	Combination	Use point-to-point terminal strip, if minor items are mostly 2-lead devices; use p-to-p plus perforated board, if minor items include more complex devices
Moderate	Mostly major components	Point-to-point	None
Complex—over 30 components	Mostly minor components	Perforated board, PC, or combination	Some circuits can be divided so 2-lead devices are wired with terminal-strip technique; rest of circuitry either perforated board or point-to-point using bolt-on terminal strips as wiring points
Complex	Equal mix of major and minor components	Combination	Point-to-point and either terminal strip or perforated board or PC, depending on kinds of minor components
Complex	Mostly major components	Point-to-point and/or combination	If more than 10 minor components, use perforated board or terminal strip to hold them

diagram was working for visual clarity, not electronic efficiency, and so symbol placement on paper may not reflect ideal electronic layout principles.

For example, when wiring an audio amplifier circuit, it is usually necessary to keep interconnecting wires—wires linking the various components—as short as possible, to minimize the risk of noise or interference pickup. But the schematic diagram of the same amplifier may show luxuriantly long leads connecting the components. If you copy the diagram in your layout, the circuit will probably not function properly.

If the business of parts layout seems mysterious, it's perfectly natural; like most portions of the design field, it's an acquired skill. Think of it in the same class as knowing how to load your car's trunk before you go on a long trip. Getting everything in the proper place seems difficult at first.

There are several general rules, and dozens of recommended techniques, most of which don't belong in a book like this. We'll discuss the important ones shortly, but here let's introduce the principle of circuit breadboarding, the most sensible approach to parts layout, and one of the most powerful project-planning techniques I can recommend.

A breadboard circuit is nothing more than a temporarily wired circuit; the components are connected—not soldered—together (with solderless connectors of some sort).

Traditionally, breadboarding has always been one of the circuit designer's bag of tricks. The technique lets him add components, or change component values, quickly, in order to arrive at desired circuit performance. In the same way, it can help you plan a circuit layout by enabling you to move components around until you've created a neat, functional layout. The process is somewhat similiar to the dry-run assembly method often used by cabinetmakers: The individual wooden sections are fitted together without glue, to make sure everything goes in place properly.

The name breadboard springs from the appearance of early temporary wiring systems—they were flat wooden boards equipped with metal clips to hold and connect wires and component leads. You can buy or build a more modern version that uses a perforated phenolic board instead of wood, and spring-loaded push-in terminals instead of clips.

By and large, the important elements of good parts layout are simply common-sense rules whose importance is obvious. Unfortunately, it is rarely possible to have every element satisfied; you often must compromise on one or two points.

• Try to keep input components well separated from output components. The idea here is to minimize the possibility of interaction between the input and output of a circuit. The concept of input and output is usually applied only to amplifier and signal processing circuits. However, a bit of thought will prove that almost every circuit has a "start" and a "finish," or a "front" and a "back," or some other configuration that resembles an input-output pattern.

• Keep AC power supply components (if present) well away from other circuitry—especially input or front-end components.

• Arrange components so that interconnecting wiring is as short as possible. Generally, longer-than-necessary wiring is not desirable in any kind of circuit, and in certain circuits can affect performance.

• Try to arrange heavy major components for good mechanical balance.

• Plan parts layout so that interconnecting wiring makes as few crossovers (nonconnecting intersections) as possible. Here, the schematic diagram is often a useful guide, since most draftsmen try to plot schematics with minimum crossovers.

• Keep "chassis ground" points as close together as possible. The whole concept of chassis grounds springs from the once almost universal practice of wiring electronic circuits on metal chassis. In many circuits, there is a common "return" path that links many components, and the metal chassis itself can be used as the conductor for this path by soldering the appropriate leads to "ground."

This scheme can be brought up to date for use with other wiring techniques by planning a common "ground bus" somewhere in the parts layout. All grounded leads are connected directly to this bus.

An important precaution, though, is to keep all ground connections physically close together. This is because even a short length of chassis surface or bus wire has a tiny—but measurable—electrical resistance. And one effect of establishing wide-apart ground points is to place these tiny resistances in series with the grounded components. In many kinds of circuits—especially high-gain amplifier and r-f circuitry—these "ground loops" can interfere with proper circuit operation.

• Keep heat-producing components away from heat-sensitive components, and try to position heat-producing components so that they can dissipate heat quickly and efficiently. (Don't, for example, box a power resistor into a crowded, unventilated corner.)

• Allow suitable free space around high-voltage circuit points to preclude "arcing." As a rule, always try to leave free space between leads and components. Don't count on insulation to prevent short circuits if you have the room to include breathing space.

Four Easy-To-Build Projects To Get You Started

Now that you know how to work with electronics in *theory*, let's get to the *practice*: Here we will discuss the design, operation, and construction of four easy-to-build electronics projects: a super-sensitive photocell relay; a burglar alarm circuit for your car; a power-tool speed control circuit for your workshop; and a general-purpose electronic "time delay" power switch.

Each of the projects is a useful, practical device that you will be proud to build and own. But what's more, if you build all four, you'll go a long way towards developing your ability to work with electronics. Here's why:

• Slightly different wiring techniques are used in the different projects so that you'll work with point-to-point wiring, 2½-dimensional wiring (on a perforated phenolic chassis board), and military style wiring (on a terminal strip).

• A wide variety of components is used in the four projects, including some of the latest semiconductor devices. All were selected because of their ruggedness—both electrically and mechanically—to minimize the chances of damage due to "beginner's error." *Note*: If you try hard enough, though, you will be able to damage these components!

• The four circuits are not critical with regard to parts placement or chassis layout. Thus, you can design your own projects from scratch, if you wish, rather than follow the chassis layouts shown in the photographs.

All four projects are relatively inexpensive (all cost less than $15) and can be built with the basic "toolbox" of mechanical, electrical wiring, and metalworking tools described in Chapter 3. And any of the projects is simple enough to build in less than six working hours—even for a beginner.

All components used are standard items on the shelves of most electronic supply houses. If your local electronics equipment supplier does not stock any of the parts, write to one of the following distributors. All of these specialize in mail-order operation, and will send you a complete catalog of components they stock:

Allied Radio Corporation
100 N. Western Avenue
Chicago, Illinois 60680

Radio Shack
730 Commonwealth Avenue
Boston, Massachusetts 02215

Lafayette Radio
Electronics Corporation
111 Jericho Turnpike
Syosset, New York 11791

Burstein-Applebee Company
1012 McGee
Kansas City, Missouri 64106

Olson Electronics, Incorporated
404 S. Forge Street
Akron, Ohio 44308

A Few Helpful Hints

You'll find all of the following suggestions in different parts of this book. But, since they are important, we'll say them again:

- Have all required parts on hand *before* you start working on any project.
- Do not use substitute parts unless you are sure that they are equivalent to the components specified in the parts lists.
- Double-check the polarity of all electrolytic capacitors and diodes before you solder them in place. Similarly, double-check the lead coding of all multi-lead semiconductor devices.
- Work quickly, with a hot iron, when you solder the leads or terminals of any semiconductor component. As a safety precaution—when you work on your first few projects—clamp a heat sink (see Chapter 3) onto each semiconductor lead before you solder it in place.
- Be especially careful when you wire leads or components that will be connected to the power line when the finished project is plugged in. An error here could blow fuses (as well as ruin components) or, possibly, create a shock hazard.
- Never connect an AC-powered project to the power line when its case and/or cabinet is open. Similarly, never open the case and/or cabinet while the power cord is plugged into a live AC outlet.

A POWER-TOOL SPEED CONTROL FOR YOUR WORKSHOP

As any workshop "pro" will tell you, the operating speed of a portable power tool—such as an electric drill or a saber saw—should be tailored to the kind of material you are working. That's why this simple—but effective—power-tool speed control will become a standard tool in your workshop. At the twist

The power-tool speed control is small enough to fit in an odd corner of even the most crowded workbench. Note the use of a three-prong (grounded) panel-mount receptacle, and (not visible) a three-wire (grounded) power cord. The ground pin of the receptacle and the ground wire of the cord must be securely wired to the aluminum minibox to insure safe use of power tools with metal housings. If you modify the parts layout, be sure to retain the location of the output receptacle on the front of the minibox, as shown. Top, rear panel, or side panel locations are not suitable because a pull on the power-tool cord will tend to tip or turn the controller. Use a large-size knob on S1's shaft (speed-adjust control); this makes small speed adjustments easier.

---

of a knob it will vary the operating speed of any series universal (brush type) motor that draws 400 watts or less. This type of motor is used in most drills, hand grinders, saber saws, and sewing machines. *Note:* Do not use this control with other types of motors—check the specification placard on any tool before you connect it.

The controller uses a conventional half-wave SCR phase-control circuit, as shown in the schematic diagram. In simple terms, that means that silicon-controlled rectifier, SCR, serves as a fast-acting electronic switch that controls the percentage of time during each cycle of alternating current that the tool is connected to the power line. The higher the percentage, the faster the tool motor turns. You set the percentage—and hence the speed—by adjusting potentiometer R1.

The circuit has the ability to maintain a roughly constant speed over considerable variations in tool load. For example, a drill will keep turning even if the bit enters a denser material.

Depending upon the type of tool, its motor, and load, the circuit will vary speed over approximately a 3-to-1 range. With most tools, the "maximum" speed setting of R1 (full clockwise) will produce 70 percent maximum speed. The control is bypassed, and the tool works normally at full speed, when switch S1 if flipped to the "bypass" position.

Building it. The speed control is built into a small aluminum minibox; the internal wiring is point-to-point. Keep these factors in mind as you work:

• You must use three-wire grounded line cord and panel mount socket to prevent a possible shock hazard caused by the internal breakdown of insulation within the power tool. This is possible. That's why most power tools (except double-insulated types) are equipped with three-prong plugs.

• Be sure that "bypass" switch S1 is rated to carry at least 5 amps AC.

• As shown in the photos, SCR is mounted on an extruded aluminum heat-sink. When you mount the component, discard the insulating washers supplied with the SCR. Apply a thin coat of heat-sink compound to the surface of the case that

---

PARTS LIST FOR POWER-TOOL SPEED CONTROL

R1—500-ohm, 2-watt, carbon potentiometer
R2—2,500-ohm, 5-watt power resistor
R3—1,500-ohm, ½-watt, carbon resistor
SCR—Motorola HEP 302 silicon-controlled rectifier
D1, D2—Motorola HEP 162 silicon rectifier
S—SPST toggle switch; 5-amp rating
SO—3-prong, panel-mount AC receptacle
PL—3-wire line cord; at least 5-amp rating
Misc.—3" x 4" x 5" aluminum minibox; 3½" x 1½" aluminum heatsink (or any similarly sized unit); heatsink compound; terminal strip; knob; grommet

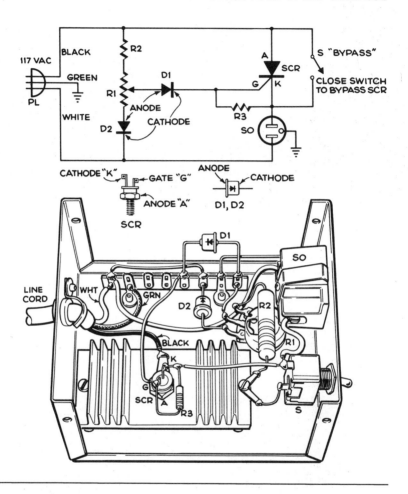

will contact the top of the heat sink, and mount the unit in place with the nut supplied. Do not "overtighten" the nut.

• *Note:* Because the SCR case is the device's "anode" connection, the heat sink will be electrically "hot" when power is applied. Thus, the sink must not touch any other components, or the aluminum minibox. Support the sink away from

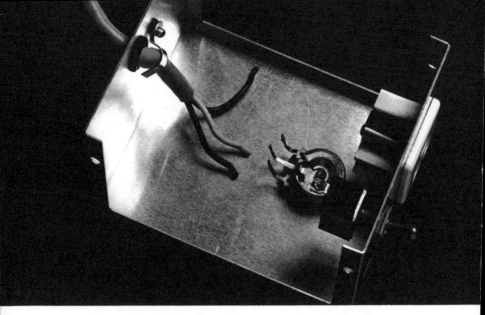

Clamp the thick power cord in place against the minibox wall before the heat sink and minor components are installed. Note that the bypass toggle switch, S1, is a heavy-duty unit capable of switching 5 amps. Conventional toggle switches are rated for 2 or 3 amps. Specify a high-current switch when you order.

The completely wired controller looks more complex than it really is. Actually, a top view is misleading since different components are positioned at different levels within the minibox. Resistor R3 must be wired directly between the SCR's gate and cathode terminals as shown. Note that power resistor R2 is positioned away from other components.

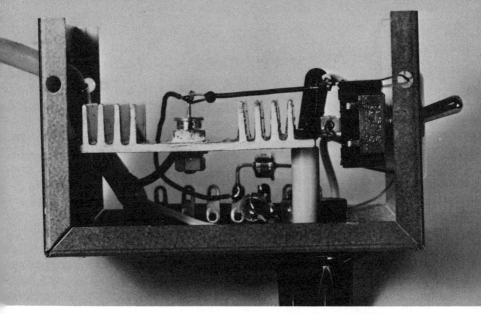

A side view shows multi-layer parts placement. Note that the extruded aluminum heat sink is supported away from the minibox surface by two ceramic insulating spacers. Since this device will be moved around a good deal, keep leads short and taut to prevent movement of wires that could cause internal short circuits.

A closeup view of the terminal-lug strip detailing the wiring of the two silicon rectifiers. Note that the strip has been fastened to the minibox with pop rivets to insure a tight mounting. Note also that no elements of the circuit are grounded to the minibox, except the ground connections of the power-cord and panel-mount receptacle, to prevent short circuits.

the minibox on *ceramic* insulating spacers. Be sure that no direct electrical connection exists between the sink and minibox.

• Use 16-gauge (or larger) wire to make all connections among the line cord, panel-mount socket, and SCR.

• As shown in the photos, clamp the line cord in place against the inside of the minibox with a cable clamp to act as a strain relief, to prevent a tug on the cord from yanking an internal connection loose.

• Position power resistor R1 away from other components; this unit gets hot when the circuit is in operation.

• Make sure that the diode bodies and other component leads cannot short-circuit against the "bottom" half of the minibox when it is installed.

Operating hints. Do not keep the controller connected to the AC power line when it is not actually controlling a power tool. Remember: Do not use the device to control a tool that draws over 400 watts AC.

A BURGLAR ALARM FOR YOUR CAR

This simple gadget will protect your car from all but the most determined auto thief. The instant he forces open the door, the circuit springs into operation and sounds your car's horn. The horn will keep blaring even if the door is closed—until you switch the alarm off by turning a key switch. If you wish, you can extend the alarm's protection to your car's trunk and engine compartment—the horn will sound if either is opened.

The circuit, as shown in the schematic diagram, uses a relay,

The automotive burglar alarm is a neat package that you can tuck into an out-of-the-way spot under your car's dashboard—it should not be installed in the engine compartment. Note that crimp-on terminals have been used to join circuit wiring to the barrier-type terminal strip. When you mount the alarm *be sure that the aluminum minibox is electrically connected to the car ground.* Run a ground wire between the minibox and the engine block if you mount the unit on an insulating surface.

RL, as a switch that turns on the horn when it is activated. RL is controlled by power transistor Q.

In operation, the base terminal of Q is connected, via resistor R1, to your car's dome light (or courtesy light) circuit. Thus, the car's dome light switches—which automatically operate when the doors are opened—serve as triggers for the alarm.

When a door is opened, and the dome light circuit is activated, battery voltage is applied to R, and hence to the base of Q. A small current flows through the base leads, which, in turn, causes a much larger current to flow through the transistor's collector circuit, activating relay RL. One pair of RL's contacts turn on the horn, while the other pair act as a "lock" that keeps RL activated even if current flow through Q stops (this will happen if the door is closed).

The only way to silence the horn is to turn off S1, cutting

BURGLAR ALARM

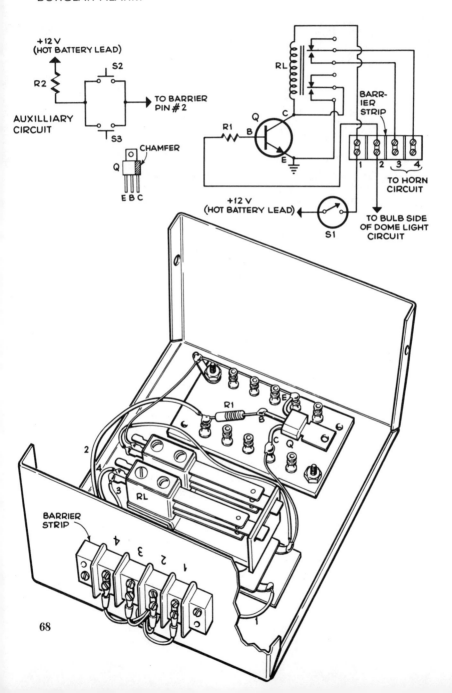

+12 V
(HOT BATTERY LEAD)

R2

AUXILLIARY
CIRCUIT

S2

S3

TO BARRIER
PIN #2

CHAMFER

Q

E B C

RL

BARR-
IER
STRIP

R1

Q

B

C

E

BARRIER
STRIP

1 2 3 4

TO HORN
CIRCUIT

+12 V
(HOT BATTERY LEAD)

S1

TO BULB SIDE
OF DOME LIGHT
CIRCUIT

R1

E

B

C Q

2

4

3

RL

BARRIER
STRIP

4 3 2 1

R1, R2—1,200-ohm, ½-watt, carbon resistor
Q—G.E. D28D1 N.P.N. power transistor
RL—Guardian electric series 200 relay; DPDT 10-amp contacts;
12-volt DC coil
Misc.—3″ x 4″ x 5″ aluminum minibox; 4-terminal barrier-type
terminal strip; section of phenolic board terminal strip; grommet
S1, S2, S3—See text

---

off power to the relay. Switch S1 is, of course, the alarm's master control. You turn on S1 after you've left the car, and all doors are closed. This "sets" the alarm. S1 can be a key-operated "lock switch" (a conventional ignition switch is fine) mounted outside the car, through a fender or door post, or it can be an ordinary toggle switch mounted in a concealed (but easy to reach) spot outside the car, say under the bumper, or behind a license plate bracket.

The accessory trigger switch circuit (shown on the diagram) is necessary if you choose to wire the car's hood and trunk lid. S2 and S3 can be any kind of spring-operated, normally open, pushbutton switches (replacement dome-light control switches are okay). Mount them, in homemade brackets, so that they turn "on" when the hood and trunk lids are lifted.

Building it. The alarm is built into a small aluminum mini-box. Since shock and vibration are problems for any equipment installed in a car, "military-type" construction (on a piece of phenolic terminal strip) is recommended. Here are the points to keep in mind as you work:

• Component values are different for 6- and 12-volt ignition systems. Note that a PNP transistor is used instead of an NPN transistor (as shown in the schematic) if the alarm is installed in a Positive Ground ignition system vehicle—the circuit as shown is designed for a Negative Ground ignition system (found in all recent American cars, and most foreign models). Check with your mechanic if you aren't sure of the type of system used in your car.

A simple, uncluttered parts layout (shown before wiring is begun) helps to insure reliability. Note that the terminal strip must be supported above the minibox surface because the base of each terminal protrudes slightly below the bottom of the insulating board.

Keep wiring direct and taut to prevent road shock and vibration from shifting wires against the aluminum surface and developing a short circuit. Note that a single ground lug—placed under one of the terminal-board mounting screws—serves as the central ground point for the circuit. Make sure that this lug makes a good electrical connection with the minibox.

The relay contact terminals are tightly spaced, so be careful when you solder leads in place. Be especially neat when you wire up the contacts that will control the horn. Keep the soldered terminals far apart. Remember that an accidental short circuit—say caused by a rough road bouncing close-packed terminals together—will sound your horn.

Be gentle with the transistor's leads when you bend them apart to fit the spacing of the terminal strip; a miscalculated yank may damage the plastic case, or break a lead off. Note the transistor's metal tab is electrically connected to the collector lead. Thus, position it clear of the minibox walls and mounting screws.

● Use 12-gauge wire to connect the circuit to your car's battery *and* horn. Use 18- or 20-gauge wire to connect the alarm to the dome light circuit and/or the accessory trigger switch circuit.

● Mount the completed alarm inside the car, under the dashboard—the minibox doesn't offer sufficient protection to install the unit under the hood.

● Make sure all connections to the barrier-type terminal strip are secure. If possible, these should be made with solderless connector lugs crimped onto the wires with a crimping tool.

Operating hints. Make it a habit to turn on the alarm whenever you leave your car unattended. Car thieves pick the oddest times to steal cars. But remember you've got an alarm. It's embarrasing to forget and then open a door.

AN ALL-PURPOSE TIME-DELAY POWER SWITCH.

Think of this circuit as a power switch with a brain. Turn it on, and a short time later it automatically turns itself off. You can adjust the time delay by turning a knob. By selecting different timing component values you can tailor your timer to delay anywhere from a few seconds to a few minutes.

What can you do with the device? Here are a few suggestions; you'll probably think of many more:

Automatic garage light or night light. Flip the switch, and your garage light or hall light will come on, and stay on long enough for you to get into your car and pull out of the

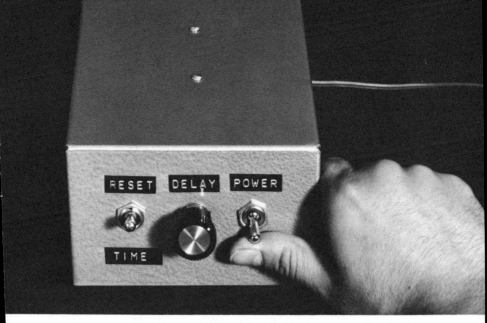

The time-delay power switch can control almost any electrically powered device that draws less than 200 watts of AC power. If you choose, you can add a time scale underneath the delay adjust-control knob, to produce a precision time-delay suitable for tasks such as controlling the "on" time of an enlarger in a darkroom.

garage, or long enough for you to walk down a hallway. Then the light will switch off—automatically.

Photographic enlarger timer. The circuit can control the "on" time of an enlarger with great precision. The timer's accuracy depends solely on how carefully you calibrate the time-delay adjust dial.

Automatic process timer. The device can control the "on" time of any small motor, machine, or circuit. The only requirement is that the process not draw more than 200 watts AC.

The circuit, as shown in the schematic diagram, has a sensitive silicon-controlled rectifier, SCR, as its heart. SCR acts as an electronic on-off switch that controls the flow of current through the full-wave bridge circuit made up of four silicon rectifiers, D1 through D4. When the SCR is "on" (in a conducting state) full-wave AC current can flow through the load (the lamp or device) connected to the panel-mount outlet, and the load works normally.

When the SCR is "off" (in a non-conducting state), no current can flow through the load. It is effectively turned off.

The state of the SCR is determined by the DC voltage applied to its gate terminal. This phenomenon is the basis of the timer's operation.

When function switch S2 is in the "reset" position, electrolytic capacitor C1 is charged to approximately 200 volts DC. When the switch is flipped to the "time" position, the charged capacitor is connected—via variable resistance R1—to the SCR's gate. Instantly, the SCR turns "on" and permits current flow through the load. The capacitor discharges slowly through the gate lead.

The SCR will remain "on" until C discharges to too-low a voltage to trigger the SCR's gate—and the SCR will turn "off," cutting off current flow through the load.

Clearly, the length of time that the load remains turned "on" depends on the length of time required for C to discharge. This, in turn, depends on the values of C and R1. The table in the parts list gives several pairs of values for these components that will let you choose the overall timing range you prefer. Within each range, the delay time can be adjusted

---

PARTS LIST FOR TIME-DELAY POWER SWITCH

R1—See table, page 78
R2, R4—4,700-ohm, ½-watt, carbon resistor
R3—3,300-ohm, ½-watt, carbon resistor
C—See table
SCR—G.E. C106B1 silicon-controlled rectifier
D1, D2, D3, D4—Motorola HEP 162 silicon rectifier
S1—SPST toggle switch
S2—DPDT toggle switch
Fuse—2-amp fuse in panel-mount holder
PL—Line cord with AC power plug
SO—Panel-mount AC socket
Misc.—5" x 7" x 3" aluminum minibox; 4" x 3½" piece of perforated phenolic chassis board; push-in terminals; copper strip; knob

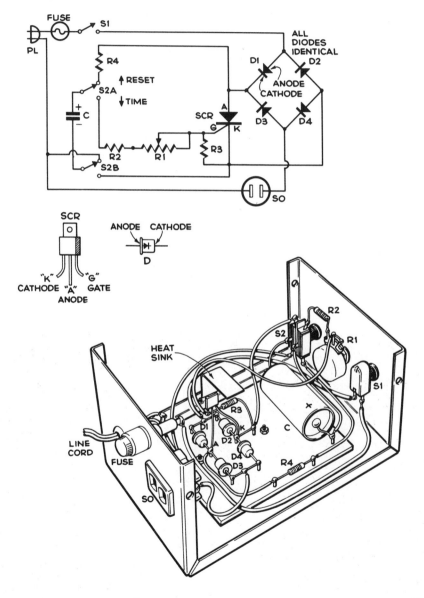

between the listed "long" and "short" values by adjusting R1.

Building it. The time-delay power switch circuit is housed in an aluminum minibox. All minor components are mounted on a perforated phenolic board; point-to-point wiring is employed. Here are the key construction points to watch:

• The SCR requires a heat sink. This component is physically identical (not electrically, though) with the power transistor used in the photocell relay. Install a similar heat sink on the tab.

• The capacitor, C, illustrated in the photos, is a 100-mfd unit. Larger value capacitors, as listed in the timing component table, will be physically larger, and may require a different parts placement fo fit within the case.

• Use 16-gauge wire to interconnect the line cord, panel-mount AC socket, and SCR.

The pre-wired perforated phenolic board (mount and interconnect its parts before you install it in the minibox) nestles below controls and other major components. Note that the circuit board must be suspended away from the minibox surface, since the lower ends of the push-in terminals protrude through the board. Also note the cable clamp used to hold the incoming power line cord against the minibox wall, to provide strain relief.

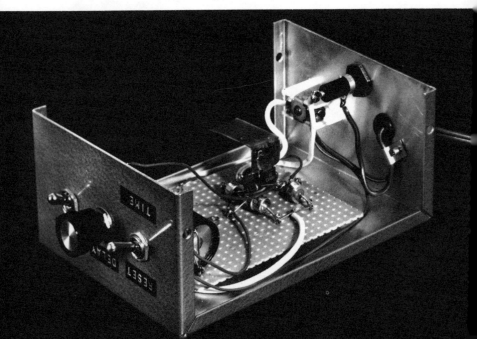

If a physically larger electrolytic capacitor is used instead of the 100-mfd unit pictured (because of a different timing range selection), you may have to modify the parts layout. There is, however, ample space within the case to handle the largest capacitor listed in the timing-range table.

The cathode and anode leads of the SCR are soldered directly to the appropriate junction points of the four-rectifier bridge; resistor R3 is connected directly between the SCR gate lead and the central "common" terminal lug (this lug is *not* connected to the case; *no circuit elements are grounded to the minibox*). Note that the heat sink soldered to the SCR's tab is bent away from other components. This is important since it is electrically connected to the SCR's anode terminal.

TIME-DELAY TABLE

All capacitors are 250-volt DC rated units.
All R1's are 2-watt carbon potentiometers.

Approximate Timing Range (may vary by factor of 2)	R1	C1
½ sec to 15 sec	100 K	50 mtd
3 sec to 60 sec	250 K	60 mtd
5 sec to 90 sec	250 K	100 mtd
10 sec to 240 sec	250 K	500 mtd

- Position all parts and wires (including the SCR's heat sink) so that none are in the way of the minibox "bottom" when it is installed.
- For accurate selection of delay times, fashion a timing scale to fit under R1's control knob (stiff paper is okay). Calibrate the scale by measuring the delay produced by different settings of R1. Use a stopwatch if possible.
- The fuse must be a 2-amp, fast-blow unit to protect the SCR. Remember, 200 watts is the maximum permissible load current drain.

A SUPER-SENSITIVE PHOTOCELL RELAY

Here's an eagle-eyed "electronic eye" that will watch over any door or window in your home. Its built-in buzzer will sound off whenever anything—man or beast—passes in front of its ever-watchful cadmium-selenide photoconductive cell (PC on the schematic diagram).

In many applications, the device won't require an auxiliary light source to cast a "light beam." Unlike most other photo-relay circuits, this one is

The photoelectric relay is a self-contained unit, including a signal buzzer mounted on the front. Note that the sensitivity control is a screwdriver-adjust potentiometer, not equipped with a knob. This control will rarely need adjustment after the unit is installed aside a door or window.

sensitive enough to work with ambient room light.

As shown on the schematic diagram, photocell PC is wired into the bias circuit of transistor Q (this is a special high-gain "Darlington amplifier" transistor). As long as light strikes the photocell's sensitive front surface, its electrical resistance is low, and a low bias voltage is applied to Q's base lead. However, when PC's surface is darkened—say by the shadow cast by someone walking in front of the cell—its resistance increases. Now, a larger voltage is applied to Q's base. In turn, this causes a large current flow through Q's collector circuit, and hence through relay RL's coil, which is wired in series with the transistor. RL "pulls in," and its contacts act as a switch that turns the buzzer on.

When PC's surface is illuminated again, its resistance drops, Q stops conducting, the relay "drops out" (opening its contacts), and the buzzer stops buzzing.

A second pair of contacts on RL can be used to control another signaling device at a remote location, if desired.

Building it. The device is housed in a long, thin, aluminum minibox. As shown in the photos, point-to-point wiring is used. Here are the key points to keep in mind as you work:

• Be sure you position RL, potentiometer R2 ("sensitivity" control), and electrolytic capacitor C, so that their terminals and/or leads don't touch the "bottom" of the minibox when it is installed in the "top" half.

• The power transistor, Q, must have a heat sink. Make one by soldering a small piece of copper (about 1" by ½") to the transistor's tab.

• The photocell is held in place by cementing it to a small angle bracket. Be sure you enlarge the bracket's hole (with a reamer) before you epoxy the cell in place so that the cell's front surface has a normal field of view.

• If you plan to use the extra set of contacts on RL, drill an additional hole adjacent to the power-cord hole, and install a rubber grommet. Route connecting leads through this hole.

• Before you close the minibox, make sure that the "bottom" half doesn't touch Q's heat sink, or cut into any wires running along the inner surface of the "top" half.

Operating hints. The device won't need an auxiliary light source if ambient room light is sufficient, or if the relay is

---

PARTS LIST FOR PHOTO RELAY

R1—250,000-ohm, 2-watt, carbon potentiometer
R2—3,300-ohm, ½-watt, carbon resistor
C—100 mtd, 25-volt electrolytic capacitor
Q—G.E. D28C1 Darlington amplifier transistor
D—Motorola HEP 156 silicon diode
T—Stancor p-6465 6.3-volt AC filament transformer
PC—Clairex CL603A photocell
RL—Guardian electric series 200 relay. DPDT contacts; 6-volt DC coil
PI—Neon-bulb pilot-light assembly with built-in resistor
Buzzer—6-volt AC buzzer
S—SPST toggle switch
PL—LINE CORD WITH AC plug
Misc.—Terminal strips; 12" x 2½" aluminum minibox; copper strip; grommet.

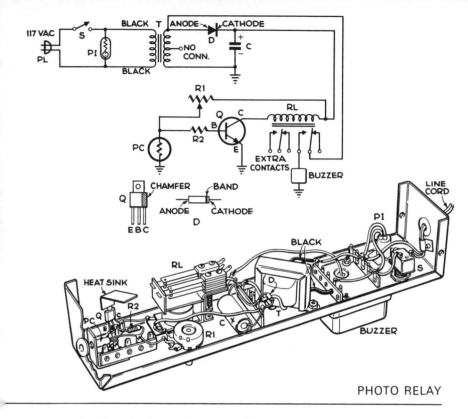

PHOTO RELAY

placed such that PC is pointed at a bright light within the room (don't count on sunlight, since the circuit will be fooled by a cloudy day).

To adjust the sensitivity control, first turn R2 full counter-clockwise. The buzzer will start buzzing. After you have made sure that a normal light level is falling on PC's surface (no one passing in front of the cell), back off on R2's setting until the buzzer is completely silent. Now, any interruption of the light striking PC will trigger the buzzer.

If ambient room light is not sufficiently bright, the buzzer will keep sounding regardless of the setting of R2. In that case, place a suitable light source opposite the photocell, on the far side of the door or window to be protected. Almost any kind of light source will work, including high-intensity lamps, a bare, low-wattage light bulb in a socket, or a light source especially designed for use with photocells (these are sold by most supply houses).

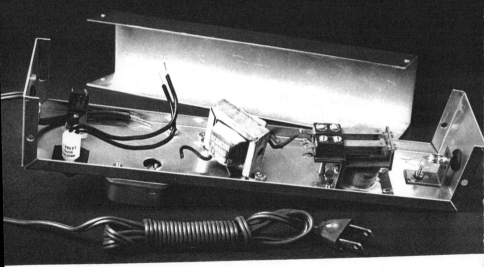

The long, thin minibox leads to a long, thin parts layout, shown here prior to wiring in minor components, or installing terminal-lug strips needed for interconnecting wires. Note the ½-inch hole above the buzzer: This is an access for the buzzer connecting wire. Experiment with parts placement carefully before you drill any mounting holes. Be sure that the pilot lamp assembly, power switch, transformer, and relay are positioned so that they won't interfere with the minibox bottom when it is slipped in place.

The cadmium selenide photocell is mounted by epoxy-cementing it to a small angle bracket. Note that the light-sensitive front surface of the cell must look through the hole in the bracket, and line up with the rubber-grommetted window drilled through the end of the minibox. Be sure you apply epoxy only around the forward edge of the cell—not across its front face.

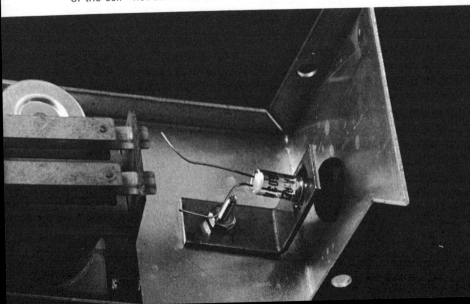

The heat sink installed on transistor Q is a small piece of copper soldered to its tab. Note that this tab is electrically connected to the transistor's collector lead, so be sure that the copper is well away from the minibox walls and other components. Also note the positioning of capacitor C; this should be centered between the minibox lips so that there is no danger of its positive leads touching the grounded chassis. If you do not plan to use the extra pair of relay contacts (as described in the text), connect the buzzer to the innermost set of contacts; leave the outside pair vacant.

The completely wired circuit is neat and trim. Because the minibox is used as circuit ground, make sure all grounded leads are well soldered to terminal lugs that are solidly connected to the box. Be sure you've used lock washers on all screws that hold grounded lugs against the chassis. Note that a small cable clamp is used to hold the incoming line cord against the minibox, and provide strain relief. Be sure that no wires are placed where they will be pinched by the minibox bottom when it is slipped in place.

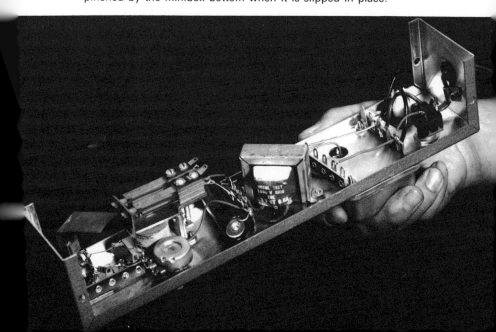

The Electronics Workshop and Tools

With a place to work and a set of tools to work with, you are ready to tackle your first job. A kitchen table makes an excellent work surface for building a ready-to-wire kit, or cobbling together a simple electronic gadget, as any apartment-dwelling electronics buff will tell you. Your woodworking bench, if you have one, is fine, too. Tools? A basic—but adequate—assortment costs about $40. Many of the tools you will use most often—screwdrivers, pliers, wrenches—are probably part of your odd-job toolbox right now.

Of course, the ideal work area, if you have the space for it, is the specially designed electronics workbench described in this chapter. And the ideal electronics toolbox contains all of the tools you will need to do the wide variety of jobs involved in building many electronics projects: metalcrafting, mechanical assembly, electrical wiring, soldering, and, occasionally, woodworking.

The Electronics Workbench

This is really two benches in one: a metalworking bench combined with an electrical-wiring table. A typical electronics project starts on the metalworking side of the bench as you drill and punch the necessary holes and openings in the project's chassis and cabinet. Then, work shifts to the electrical-wiring table where you mount the major electronic components (the large ones that are bolted directly to the chassis or cabinet) and solder in place the minor components (small parts that are held in position by their leads when you solder them to terminal strips or other soldering points).

As the drawing indicates, the wiring table should be only

Electronics workbench consists of metalworking bench on one side of storage cabinet and wiring bench on the other. Wiring bench should be covered with a light-colored formica so top is heat-resistant and small parts can easily be seen. Tools are hung on perforated hardboard panels within easy reach.

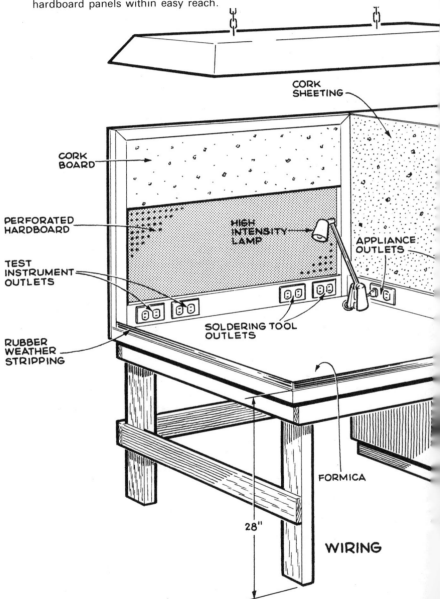

CORK
SHEETING

CORK
BOARD

PERFORATED
HARDBOARD

HIGH
INTENSITY
LAMP

APPLIANCE
OUTLETS

TEST
INSTRUMENT
OUTLETS

SOLDERING TOOL
OUTLETS

RUBBER
WEATHER
STRIPPING

FORMICA

28"

WIRING

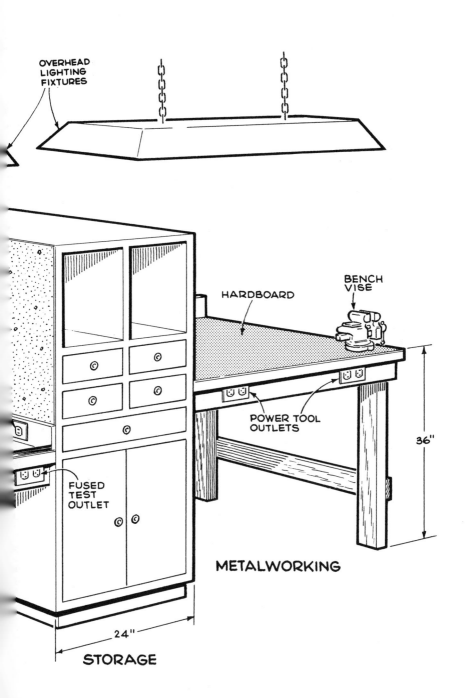

OVERHEAD
LIGHTING
FIXTURES

HARDBOARD

BENCH
VISE

POWER TOOL
OUTLETS

36"

FUSED
TEST
OUTLET

METALWORKING

24"

STORAGE

about 28 inches high—roughly 8 inches lower than the metal-working bench, which is at conventional workbench height. Since you will be seated when you do most wiring work, this arrangement assures that your elbow won't pop into the air every time you raise your soldering iron into working position.

The 2-foot-wide divider section, containing several storage drawers and cubbyholes, smooths out the height differential between the two work surfaces, and also shields the wiring table from metal chips and filings produced by the metal-working side of things.

The perforated hardboard backboard, behind the wiring table, lets you hang your wiring and soldering tools within easy reach. This is helpful, since you normally use a dozen or more tools in rapid succession as you work. Also, the backboard makes a convenient—and highly visible—place to hang test meters, or other instruments, when you take measurements or troubleshoot a chassis. Incidentally, since you will spend a good deal of time in front of the perforated board, paint it a dull, nonglare finish.

Cover the electrical wiring table with a smooth, heat-resistant surface; kitchen-cabinet-grade formica, or similar material, is fine. Choose a neutral gray or light pastel surface color. Darker colors make it difficult to spot small electronic components and tiny hardware lying on the surface. Preferably use mat-finished material. Its dull, nonreflective surface helps reduce annoying glare.

Cement a strip of rubber weather stripping—the kind designed for metal storm doors—around the wiring table's perimeter. It should stick up about ½ inch above the surface to form a guard rail that will prevent small round components and hardware from rolling off the table, yet will allow you to sweep the surface free of bits of wire and insulation simply by pressing down the flexible barrier.

Cover the side of the divider section that faces the electrical wiring table with cork sheeting, and add a foot-wide strip of cork board (cork sheeting cemented to a hardboard backing) to the top of the perforated backboard. These "bulletin-board" surfaces are a convenient place to tack up schematic and pictorial diagrams when you work.

A word about seating: An office-type swivel chair (you can buy one quite inexpensively at a used office-equipment shop) makes a perfect companion for the workbench. Choose an armless model that will roll in close to the work table, yet will allow you to reach all four corners of the wiring area without standing up. It's convenient and very comfortable. You'll appreciate this most after a lengthy wiring session.

Lighting

The success or failure of a project often depends on your ability to see what you are doing in great detail. For example, most resistors and semiconductor diodes are identified—their values indicated—by tiny adjacent color bands on their bodies. In dim light, it is rather difficult to spot subtle color differences. Or, looking at it another way, it's easy to mix up two similar appearing, but electrically different, components, and make a wiring error.

Soldering, too, requires good lighting, since you judge the quality of a solder joint by examining its surface. A good solder joint has a shiny, metallic, almost silvery sheen to it; a bad—or "cold soldered" joint connection—looks dull and grainy by comparison. These slight differences show up only under a bright, but glare-free, light.

Lighting engineers talk about illumination in terms of "footcandles" (or fc, for short)—a unit of light measurement that dates back to the early days of optics, when the flame of a sperm oil candle was the standard light source used to calibrate light-measuring instruments. One fc represents the illumination on a surface that is 1 foot away from a burning standard candle. As a rule of thumb, 100 fc is roughly 10 percent of the outdoor light level on a very cloudy day.

Industrial surveys have verified that the proper light level for "general assembly work" is about 70 fc, while the recommended light level for "fine and detailed assembly," and "color identification," is about 200 fc. In more practical terms, this means that the overall illumination on your workbench should be 70 fc, and that the illumination of a small area on the wiring table—the area where the chassis sits as you wire it—should be three times as great. Actually, your eyes respond geometrically, not linearly, to light level increases, so the 200-fc area will seem only twice as bright.

In most cases, the best way to provide the recommended illumination levels at your workbench is with two sets of lights. Let your workshop's general lighting system—including ceiling, wall, and free-standing light fixtures—provide the overall 70-fc illumination level. The additional 130 fc (to total 200 fc at the center of the wiring table) can come from one or two tabletop high-intensity (Tensor-style) lamps.

With the help of a conventional photographic exposure meter, you can determine in a few seconds whether or not your workshop's (or work area's) present lighting setup is satisfactory. Take a meter reading by holding your light meter about 2 feet away from a large sheet of white paper held horizontally at workbench height. Be careful you don't take a reading of your arm's shadow on the paper. If your meter is supplied with an illumination-value conversion table (check the instruction manual) use it to transform the meter reading into an equivalent footcandle reading. If not, set the meter's exposure index dial (ASA index dial) to ASA 100 (DIN 21), and manipulate the exposure-computer dial just as if you were planning to take a picture of the piece of paper. An Exposure Value scale (EV scale) reading of between 10.5 and 11 corresponds to an illumination level of about 70 fc; a reading of between 12.5 and 13 corresponds to 200 fc.

Electric Outlets

A good plan is to provide a receptacle for every appliance, tool, instrument, and accessory that you will use at the bench. Ideally, every outlet that will accept a hand-held power tool or relatively high-powered appliance should be of the three-prong grounding variety.

The simplest solution is to install a multi-outlet socket strip across the front of your workbench. Unfortunately, this invariably leads to a tangle of power cords all leading to the same socket strip. The best approach to planning your work-bench's wiring scheme is to consider where on the bench each of the various AC-powered appliances and devices will be used:

Soldering tools. Install two outlets; one for a continu-ous-duty soldering iron, and the other for an occasionally used soldering gun (the section on soldering tools later in this chapter explains why both tools are necessary). Locate them so that the tools can lie close to your working hand, without their power cords dangling across the wiring table. If you are right-handed,this means you should position the outlets on the right-rear corner of the wiring table.

Test instruments. Mount four outlets on the wiring table's rear edge especially for test instruments, even if you don't own any AC-powered gear right now. As your interest in electronics grows, you'll almost certainly acquire a vacuum-tube voltmeter (VTVM), and an experimental-circuit power supply. You may even eventually add an oscilloscope and a signal generator.

Power tools. Two outlets, located on the front edge of the metalworking bench, are adequate for the few power tools you are likely to use for electronics assembly: a portable power drill, a saber saw, and possibly a modelmaker's lathe.

Appliances and accessories. Two outlets, mounted on the divider section, may be enough, but you might need a dozen accessory outlets scattered across the workbench. It all de-pends on how many additional appliances and accessories you feel are necessary. You'll need a minimum of two sockets for the high-intensity lamps we mentioned earlier. But then, do you want a radio? A clock? A fan? An electric space heater? Well, you get the idea.

Test socket. For safety's sake, you should provide an inde-pendently fused and completely *isolated* outlet to power just-completed projects and circuits under test. The idea is to pre-vent a short circuit on the experimental side of things from blowing out the workshop lights, or from presenting an unex-pected shock hazard.

The fusing requirement is easy to meet. Both sides of the

power line running to the test receptacle are fused, with fast-acting, easily replaceable fuses. Achieving complete isolation is a bit more complicated: you'll need a device called an isolation transformer. Basically, this is a transformer with identical primary and secondary windings, so its output voltage is equal to the input voltage. It isolates the chassis or project you are working on from the main power line by the simple technique of eliminating any *direct* connection between the two. Your project plugs in to the transformer's secondary; the transformer's primary runs to the workbench's main junction box. The only connection between them is the transformer's magnetic field.

Why You Need an Isolation Transformer

To understand how this setup can protect you from potentially dangerous shocks, we must first investigate their source. One side of the two-wire 120-volt AC power line running through your home (actually, most homes have more than one line, but they are identical) is electrically grounded. Thus, if you were to connect an AC voltmeter between the ungrounded side of the line and a grounded metal object—a water pipe, for example—you would observe a 120-volt AC reading. And this also means that if you were accidentally to touch the ungrounded side of the line at the same time another part of your body was in contact with a grounded object, you would receive a nasty—and possibly dangerous—shock. There are several different ways that this situation can develop when you work on a piece of AC-powered gear plugged into an ordinary outlet. Carelessness usually plays an important part, but not always.

For example, suppose that a faulty component in the hi-fi kit you've just assembled accidentally short-circuits one wire of the device's power cord to its chassis. In many modern circuit designs this type of short circuit probably won't pop any fuses, but it creates a serious shock hazard. Actually, it's a shock hazard 50 percent of the time: it depends which way you insert the power plug in the outlet. If you are lucky, you will connect the grounded side of the power line to the shorted wire of the power line; if you aren't, you'll insert the plug the other way around, and connect the ungrounded—or live—side of the power line to the shorted wire. In the latter case, you've inadvertently placed your project's chassis at a potential of 120 volts AC. If you touch the chassis for any reason when its power cord is plugged in—to adjust a chassis-mounted control, for example—it's equivalent to sticking your finger into the ungrounded slit of a power outlet. Of course, you won't be shocked unless some part of you is touching a grounded object, but you are running a substantial risk. Consider: the metal chair you may be sitting on, which is in contact with a damp basement floor, may be a good ground. Or the metal electric space heater your leg brushes against is almost certainly a well-grounded object.

Another important fact to keep in mind is that all AC-DC electronic gear—including most table-model AM radios, many TV sets, and a number of simple phonographs—are designed to have one side of their power cord connected to their chassis. Although all such equipment should be fitted with a polarized power plug (the two prongs are of different size, so you are forced to insert the plug in the right direction), to insure that the chassis is always connected to the grounded side of the power line, most AC-DC devices aren't. Consequently, since there is no way of guaranteeing that the chassis won't be "live" with respect to ground, AC-DC equipment usually has outer cabinets made of plastic or wood that insulate the chassis from prying hands. Moreover, an interlock on the power cord cuts the flow of current to the device whenever the cabinet is opened.

If you attempt to service an AC-DC device, however, you no longer have the protection of either the cabinet (you must get at the chassis in order to service it) or the interlock (you must defeat it so you can observe the chassis in action). Because of this potential shock hazard, most AC-DC equipment carries explicit warnings that tell electronic novices not to attempt servicing themselves.

An isolation transformer eliminates this type of shock danger, whether you are dealing with AC-DC equipment, or other AC-powered devices that may have an accidental chassis-to-power-line short circuit. In fact, it will even protect you from your own carelessness: without an isolation transformer to guard you, you would receive a painful shock if a metal tool you were holding accidentally brushed against a "live" terminal on an AC-DC chassis (or the "live" side of the power transformer primary wiring in any other AC-powered device), and another part of your body were grounded.

With the isolation transformer between your project and the power line, there is no longer a direct connection between ground and *either* side of device's power cord. That's why it's a worthwhile addition to your electronics workbench. Choose a transformer that is rated for a 150-watt load. Mount it in one of the divider section's cubbyholes; locate the test outlet and the dual fusebox just below the front edge of the electrical wiring table.

One final point: Even if you don't build the electronics workbench, I recommend that you use an isolation transformer when you test and service your AC-powered projects and other electronic equipment. Since most transformers in the 150-watt-rating category are supplied with built-in output outlets, you can simply place one on your work table, and plug your project in. What about fusing? Just add a fused plug (containing two 2-ampere, cartridge-type fuses) to the transformer's power cord. Of course, don't connect a device that draws more than 150 watts into the transformer.

Basic Tools for Electronics Work

Your electronic toolbox contains the tools you will use for the mechanical assembly and electrical wiring of the projects and kits you build. It must include certain tools, and isn't really complete without several others. Here are the essential tools:

Screwdrivers. Electronic devices—even simple ones—invariably use a wide variety of screw fasteners: #2 or 4 setscrews in control knobs; #2 machine bolts to mount tube sockets; #4, 6, or 8 machine bolts to mount major components, terminal strips, and some chassis-mount connectors; and #6 or 8 sheet-metal and self-tapping screws to assemble

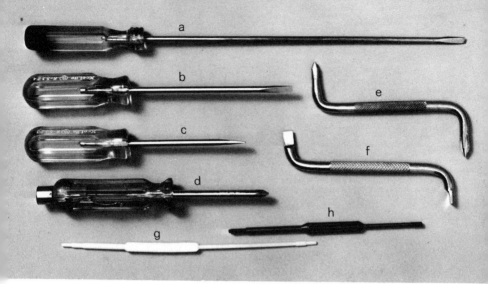

Six screwdrivers you probably won't find in your general toolbox: (a) long-shaft ⅛" standard blade; (b) ⅛" cabinet blade; (c) $\frac{3}{16}$" cabinet blade; (d) #2 Phillips head with $\frac{5}{16}$" nutdriver on handle; (e) offset Phillips head; (f) offset cabinet blade; (g and h) plastic body alignment tools used to adjust variable inductors.

chassis and cabinets. This is only a partial list. Fortunately, only three different screwdriver blade sizes will handle most of the screws and bolts you are likely to run across: ⅛ inch, ¼ inch, and $\frac{5}{16}$ inch. The ⅛- and ¼-inch drivers should have electrician's (or cabinet) blades. This style of blade has a rectangular-shaped tip, instead of the more familiar keystone shape, so it can turn screws in deep recesses and holes.

Phillips-head screws are almost never used in home-built projects, and rarely in ready-to-wire kits. However, you will find them used often in commercially built devices, especially where the designer wanted to minimize the chance of tampering by inexperienced persons. Apparently, the odd-looking star-shaped slots are supposed to frighten away the uninitiated. I've found that two Phillips screwdrivers—a #1 and a #2-size—will take care of most commonly used Phillips-head screws.

Screwdrivers of all sizes and shapes come in a bafflingly wide range of prices, and a price tag isn't always an accurate guide to quality. Remember, in a screwdriver it's the blade that counts: a good-quality blade has square, *sharp* edges that grip a screw tightly as you turn. The handle material, and the way the shaft is mounted in the handle, aren't quite so important, for you will rarely have to use substantial amounts of torque when you mount electronic components.

Longnose pliers. You will use this tool to shape the ends of component leads prior to soldering, to crimp wires around terminal points (and so form a good mechanical as well as electrical connection), and to serve as a heat sink when you solder a heat-sensitive component in place (by gripping the component's lead between the component and the solder joint). This is the wiring tool you will use most often, so buy the best quality you can afford. Select a 6- or 7-inch pair; the slenderer the noses are, the better. On top-quality pliers the jaws taper to needle-sharp points, and sit flush against each other along their full length when the pliers are closed. Two

A gentle grasp or a grip of steel. Left to right: spring-action tweezers; hemostat (locking forceps); miniature longnose pliers; 7″ longnose pliers; 7″ locking pliers.

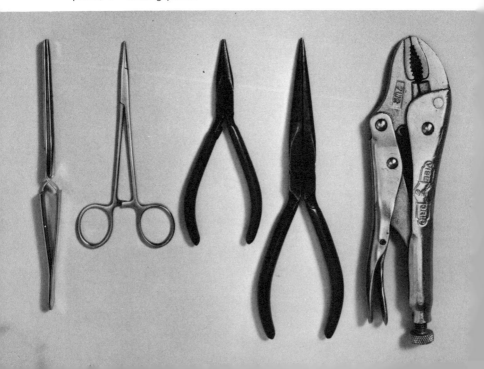

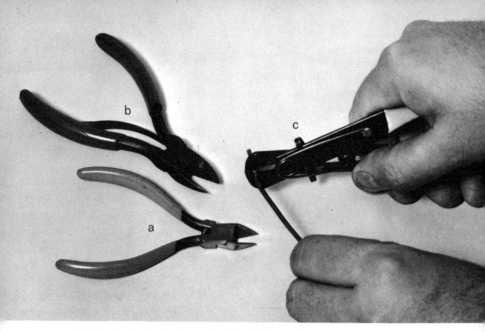

A trio of wire-handling tools: (a) miniature diagonal cutters; (b) 5″ diagonal cutters; (c) pliers-type wire strippers.

extra-cost options you don't really need, but are occasionally useful, are built-in wire cutters at the rear of the jaws, and a spring-loaded action that pops the jaws open automatically. On the other hand, an accessory you really should have is a pair of cushion-grip slip-ons for the handles. Expensive pliers are supplied with them already molded on.

Diagonal cutting pliers. These are used to cut component leads and lengths of hookup wire to approximately the proper length before wiring them in place on a chassis, and then to cut away excess leads protruding from a solder joint after the solder has hardened. A 5-inch pair provides sufficient leverage to cut the heaviest-gauge wire you are likely to use in an electronic project, yet is small enough to manipulate in the crowded quarters of a tightly packed chassis. An easy way to judge quality is to hold the cutters up to a light. Press the jaws together tightly; you should not be able to see any pinpoints of light between the two cutting edges. A word of warning: never use diagonal cutters to cut steel wire or nails—the fragile cutting edges are easily knicked.

Wire stripper. Although you can use an electrician's knife to strip the insulation off the ends of hookup wire, a wire stripper usually does the job better, and much faster. Although there are several mechanical semiautomatic stripping devices available, I've found that the simple pliers-type stripper is best of all, unless you are running a chassis-building assembly line. Most beginners find the spring-loaded, self-opening variety easiest to master. It takes a bit of skill to cut through a wire's insulation, without nicking—and weakening—the copper wire underneath. Incidentally, never use a wire stripper as a wire cutter; its cutting edges are highly sharpened to slice through plastic insulation, and are quickly dulled by cutting copper wire.

Dogbone wrench. This is a poor-man's version of a set of Spintite nutdrivers, and in most cases it works just as well. Its ten openings will handle just about every size hex nut you may be required to tighten, except one: the mounting nut for a conventional panel-mount fuse holder or pilot lamp. You'll need an additional adjustable wrench for this.

Tweezers. In those really tight quarters where both long-nose pliers and your fingers are useless, tweezers are used to bend component leads, pick up bits of wire and drops of solder, and to manipulate closely spaced components. Choose a pair about 10 inches long with needle-sharp points and moderate spring action.

Adjustable open-end wrench. Although you will find it a bit clumsy to maneuver under a chassis, an 8- or 10-inch adjustable wrench will tighten virtually every hex and square nut used in electronic devices (including the elusive fuse-holder mounting nut), and is the perfect tool for operating a moderate-sized chassis punch (see section on metalworking tools). It belongs in your toolbox, but be careful when you use it. With a wrench this long it is all too easy to apply excessive torque, and damage the component or bolt you are mounting. Buy a wrench whose adjusting worm is equipped with a locking device; manipulating an ordinary nonlocking wrench in close quarters invariably joggles the worm and changes the wrench's size.

Electrician's knife. It's similar to an ordinary pocket jack-knife, but it is equipped with a locking device that keeps the blade in use fixed rigidly in the extended position. Most electrician's knives have at least two blades: a broad, general-purpose cutting edge that is ideal for slicing through the outer plastic jacket on coaxial or multiconductor cable; and a notched wire-stripping blade. Drawing enamel-coated wire through the notch is about the easiest way I know of to scrape off the insulation.

Heat sink. This is a vital tool for handling heat-sensitive semiconductor components. For about half a dollar, you can buy a commercially made heat sink that looks like a miniature carpenter's spring clamp. Or, if your wife isn't looking, you can swipe one of her flat, spring-loaded hair-curler clips. Both work as well, and in the same way: when clipped to the lead of a component, they draw away excess heat during soldering that would otherwise reach—and possibly damage—the component. Keep in mind, though, that a heat sink is not a cure

An assortment of wrenches that will handle 95 percent of the hex nuts commonly used in electronics. Left to right: miniature and 8″ adjustable open-end wrenches; dogbone 10-opening hex-nut wrench; ¼″ nutdriver.

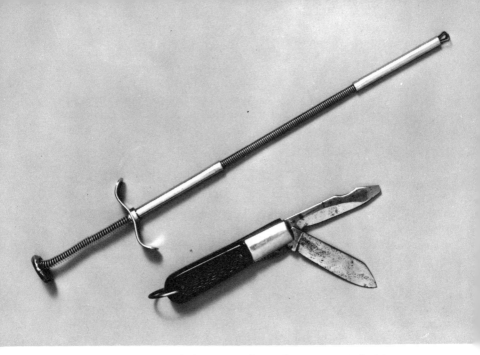

Two tools to keep handy in your toolbox: plunger-operated pickup tool (top) for retrieving small parts dropped into chassis crevices; electrician's knife with screwdriver, cutting blade, and wire stripper.

A lifesaver for overheating semiconductors: a clamp-on heat sink that will draw away damaging heat when you solder leads in place.

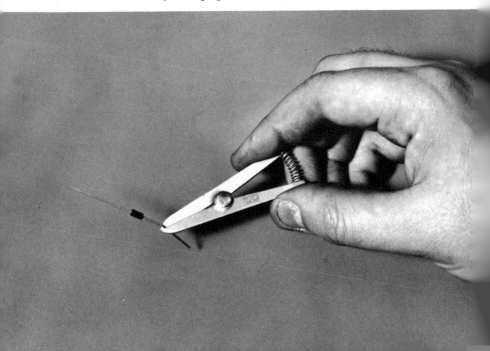

for bad soldering technique; you still must learn to solder quickly, with a small iron, when you work with delicate transistors and semiconductor diodes.

Common straight pin. Yes, an ordinary straight pin from your wife's sewing basket. There is nothing handier when it comes to "starting" a stiff transistor socket. The contact tension in a new socket is usually so great that it is impossible to insert a transistor without twisting, bending, and possibly breaking the leads. The solution? One or two gentle stabs in each of the socket's holes with your straight pin.

Allen wrenches. Treat your set of allen wrenches like an insurance policy: buy it, and put it away. Allen-head bolts and setscrews aren't very common in ready-to-wire kits, and you'll probably never go out of your way to use one in a home-built project. But if you do come across one—perhaps in a piece of commercially built gear—the only way you can loosen or tighten it is with the proper size allen wrench.

Miscellaneous. In addition to the above-mentioned tools, you'll probably find the following useful: a pair of scissors (6 to 8 inches long); a 12-inch metal rule; a pocket magnifying glass; a pair of conventional 8-inch slipjoint pliers; and a flashlight.

Special-Purpose Tools

There's no doubt that owning exactly the right tool for a job makes the job easier to do. As you work with electronics you will find yourself performing a wide variety of tasks that can be performed more efficiently by tools not included in the basic toolbox outlined above. No, these extra tools are not merely gadgets; they are special-purpose tools designed to handle occasional—and invariably, tricky—jobs with ease. All of the tools listed below are available from most electronic supply shops, including all of the large mail-order supply houses.

Special-purpose screwdrivers. An offset-blade screwdriver has a wrenchlike handle set at right angles to the blade that lets you apply enormous torque to the screw you are tightening or loosening. While this added twisting force is neither necessary nor desirable when you are working on an electronic chassis, the offset blade is ideal for manipulating a bolt in tight quarters, or one that is partially hidden by other components. You will often run into the latter situation if you are forced to replace a major component on a fully wired chassis. Most offsets have dual blades: on one side of the handle the tip is parallel to the shaft; on the other side it's at right angles. This arrangement lets you tighten a bolt in close spots where you can swing the handle only a quarter turn at a time.

In a slightly different way, a screwholding screwdriver lets you manipulate a bolt with ease in a crowded chassis. Its spring-steel jaws grip the bolt, and hold its slot on the blade's tip. Then you can insert the bolt in a hole or other location where your fingers can't reach. Once the bolt is started in the nut, you can release the jaws and use the screwholder as a conventional screwdriver. Although this type is made in both standard and Phillips form, in several sizes, I find my ¼-inch cabinet-blade model most useful.

Miniature longnose pliers and diagonal cutters. The 4-inch versions of these indispensable tools are ideal for wiring printed circuit boards, or for working in cramped quarters around a component-choked tube socket or terminal strip.

Nutdrivers. These permit you to tighten or loosen a bolt from either side of the chassis. Hollow-shaft nutdrivers are best; they let you turn a nut even if a long length of threaded bolt shaft protrudes through it. Three nutdrivers—$1/4$, $5/16$, and $11/32$ inch—will handle virtually all of the machine nuts used in electronics. The $1/4$-inch size, incidentally, is also the correct tool for turning #6 hex-head slotless metal-tapping (or sheet-metal) screws often used in commercially built equipment.

Ignition pliers. These miniature slipjoint pliers have an offset head and a relatively flat profile, giving this tool remarkable agility in tight quarters. Although originally designed to handle the small hardware used in automobile ignition systems, my pair is equally at home tightening mounting nuts on binding posts and miniature connectors.

Extra-heavy diagonal cutters. Occasionally, you may have to cut large-gauge copper "bus-bar" wire, or clip off unused lugs and terminals from tube sockets, potentiometers, connectors, and terminal strips. A 7- or 8-inch diagonal cutter is the best tool for the job.

Hemostat (also called seizers or locking forceps). Surgeons use this tool to pinch off bleeding blood vessels, but I find that a hemostat makes an ideal third hand when I solder small components in place. The device will lock its jaws around a component's lead with a tenacious, bulldog-like grip, enabling you to position the lead at precisely the desired angle as you solder it. And a hemostat is an excellent high-capacity heat sink; there's no better protection for delicate semiconductor components.

Electronic vise. There are several special-purpose mechanical supports available which are designed to hold a small chassis or printed circuit board at a convenient angle for soldering. They consist of an adjustable chassis holder linked to

The hemostat's self-locking jaws can serve as a heat sink around the lead of a semiconductor component, or grip a small part tightly when a "third hand" is needed.

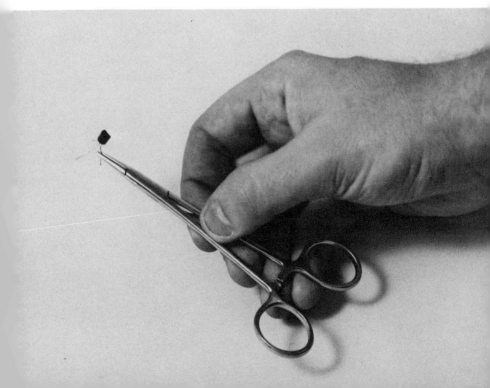

a heavy base by a lockable ball-joint. I have found that a conventional miniature clamp-on metalworking vise is a less expensive—and only slightly less versatile—substitute.

Pick-up tool. This device saves you the trouble of turning a chassis upside down, and shaking it, to get hold of a small component that has fallen into a crowded corner. It consists of moveable spring-steel jaws on one end of a long flexible shaft, and a finger-operated plunger control on the other. Press the plunger, and the jaws open; release the plunger, and they spring shut on the elusive component.

Inspection mirror. This version of your dentist's little mirror-on-a-stick lets you examine the hidden sides of crowded terminal strips for signs of a short circuit or a poorly soldered joint. Some models have built-in battery-powered illuminators.

An inspection mirror lets you look inside hidden chassis crevices to examine out-of-sight solder joints and connections.

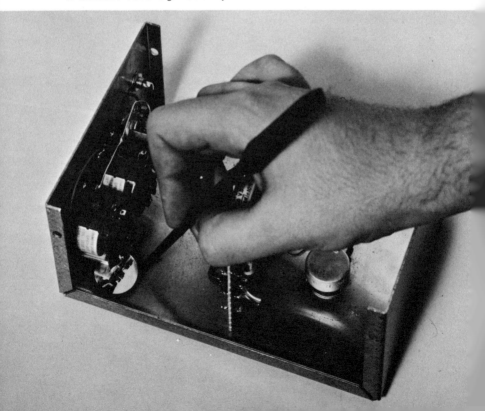

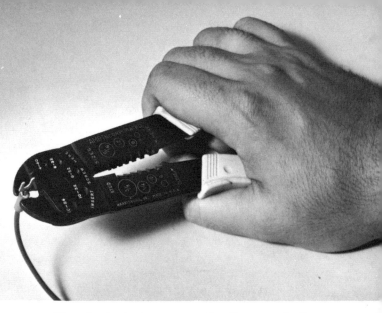

Squeeze . . . and the crimping tool compresses the slip-on terminal's metal collar around the stripped wire end, forming a *solderless* connection that has excellent mechanical strength.

Tube puller. This handy gadget helps keep the skin on your knuckles when you remove a suspect vacuum tube from a crowded chassis. Modern tube sockets are designed with very stiff pin contacts to insure a good electrical connection to the tube, and often you must exert a substantial pull to free a tube. You operate a tube puller like a pair of longnose pliers; the device's cylindrical jaws distribute the squeezing force along most of the tube's glass envelope, so there is little likelihood that you will break the tube.

Tube pin straightener. Another "insurance policy" item, and a worthwhile investment if you plan to service or build vacuum tube equipment. Its steel dummy sockets will quickly straighten any tube pins you accidentally bend when you pull a tube. Choose a model that accepts both 7- and 9-pin miniature tubes.

Crimping tool. Crimped—or solderless—connections are often used to attach terminals and connectors to the ends of connecting cables. For example, the dual conductor cable that joins a speaker system to an amplifier may have *spade-lug* terminals crimped on both ends of each conductor: the spade lugs slip under the terminal screws on the backs of the speaker and amplifier. However, most applications of this very versatile technique are electrical rather than electronic. Virtually all automotive electrical-system wiring is done with crimped connections, and so is a substantial amount of house wiring.

To make a crimped connection, you need a pliers-like device called a crimping tool. It compresses the terminal's or connector's metal collar around the stripped end of the wire to form an excellent electrical and mechanical connection. Many different shapes and types of terminals are available, with different collar sizes to fit a wide range of wire gauges. As a bonus, many crimping tools are equipped with built-in bolt cutters that enable you to shear standard bolts into odd lengths.

Riveting tool. Pop rivets are ideal fasteners for mounting terminal strips, switches, tube sockets, and other small components to metal chassis and cabinets. They are mechanically strong, and form an excellent electrical connection as well. This is important, since the mounting lugs of terminal strips and tube sockets are often used as a grounded connection to the chassis. Pop rivets are installed with an inexpensive, hand-held riveting tool.

One word of caution: don't use rivets—even the soft aluminum variety—to mount plastic or phenolic components; their mounting lugs may shatter as the rivet body expands and locks.

Labeling machine. This gadget makes professional-looking labels for your home-built projects. Just squeeze the handle and spin the letter-die wheel. The adhesive-backed plastic labels come out in an easy-to-handle strip. Use them to label control functions, tube and transistor sockets, and input and output connectors and terminal strips.

Alignment tools. You'll need suitable alignment tools when you build or service any electronic equipment that uses adjustable coils or trimming capacitors. The alignment tool's task is to allow you to turn the component's moveable core or adjustment screw, and, at the same time, to isolate the compo-

Riveting metal components to a chassis or cabinet is easily done with a hand-operated riveting tool. Use the softer aluminum rivets for fastening to aluminum surfaces. Don't try to rivet plastic or phenolic components; they may shatter under the riveting pressure.

nent from your body capacitance which could affect circuit operation. All alignment tools look pretty much the same: a slender plastic shaft topped by an odd-looking tip. There is a wide variety of tip styles available to fit the wide range of different variable components. It makes no sense to buy a complete selection; rather build up a library of styles slowly by acquiring tools as you need them. That's why I consider alignment tools as extras. However, they are necessities when you adjust variable coils and trimmers.

Metalworking Tools

Sure, it's possible to work with electronics and never build a project from scratch. But, sooner or later, most people get a creative urge to go beyond wiring a kit, to plan and execute a project entirely by themselves. This invariably means metalwork. A simple project requires, at the very least, that you drill a few mounting holes in a metal cabinet. More complex projects often involve such tasks as bending small metal brackets and cutting odd-shaped holes in metal chassis.

Not so long ago, everyone bent and formed their own metal chassis and cases. If you are so inclined—and if you own a metal brake—you can carry on this tradition. However, in these times of relatively inexpensive, beautifully finished, ready-made chassis and cabinets, it doesn't make much sense. We will say more about this in Chapter 4. The metalworking tools listed below are the ones you will need to handle the many simple drilling, cutting, and punching jobs you are likely to encounter as you transform a bare chassis and cabinet into the framework of an electronic project.

Hand drill. A simple—and venerable—crank-operated hand drill is your basic hole-making machine if you don't own a power drill. If you do, a hand drill is still essential: it will work in those circumstances where a power drill is impractical. A few examples: To drill a blind hole through an already wired chassis—you can control a hand drill with great precision, and take care that you don't drill through any components mounted under the chassis. To make a hole in a piece of fragile phenolic board (which is often used to hold minor components)—a heavy, high-speed power drill may crack the board. To drill a small hole through a painted cabinet—with a hand drill you run a smaller risk of the bit skittering across the surface and ruining the finish.

Set of drill bits. Choose "high-speed" drill bits rather than carbon steel bits, even if you don't own an electric drill; they will last considerably longer. Seven different sizes—$^1/_{16}$, $^3/_{32}$, $^1/_8$, $^5/_{32}$, $^3/_{16}$, $^7/_{32}$, and $^1/_4$ inch—make a sufficiently comprehensive set for electronics work. Also, you should have a sharp scriber and a center punch to mark the location of holes before you drill.

Tapered reamer. This tool looks like an elongated counter-sink; it is used to enlarge small drilled holes. The most useful reamer tapers from ⅛ to ½ inch in diameter. This simple device eliminates the need for a large assortment of drill bits; if you must drill an odd-sized hole, choose a slightly smaller bit from the set of seven described above, and use your reamer to enlarge the hole it makes to the desired size.

Chassis punches. Simple—and very effective—tools for making moderate to large round and square holes in sheet metal. They consist of a punch which is pulled through the sheet metal, into a matching die, by a drive bolt. To make a hole, you first drill a smaller pilot hole, the diameter of the drive bolt. Then you place the punch and die on either side of the hole, and join them by passing the drive bolt through the die and threading it into the tapped punch. As you turn the bolt with a wrench, the punch shears through the metal,

---

Basic metalworking tools, left to right: nibbling tool; chassis punch; hand-driven tapered reamer; center punch; scribing awl.

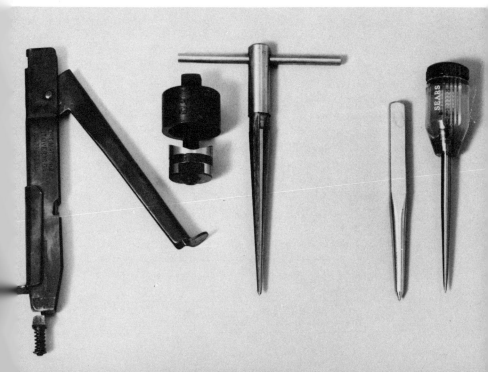

forming a clean hole. Round chassis punches are available in a wide range of diameters from ½ inch to 3 inches. The ½, ⅝, $11/16$, and 1 inch sizes are most often used. You don't really need any square punches—the few square holes you will make are easily cut with a nibbling tool.

The hand-operated nibbling tool can be used to cut odd-shaped openings in thin-gauge aluminum and steel chassis and cabinets. Start by punching a pilot hole; then use the tool to nibble away excess metal to form the desired opening.

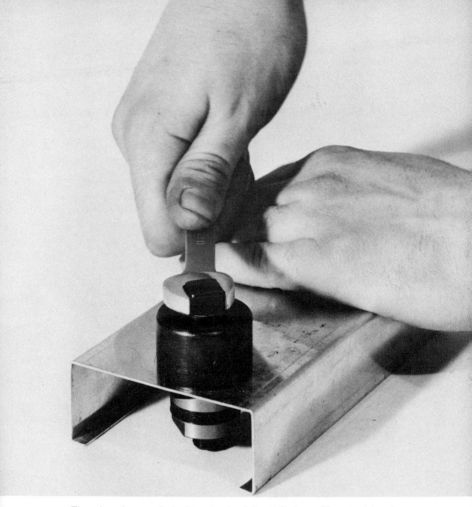

The chassis punch is the standard large-hole-making tool in electronic construction. Punches are available in a wide assortment of sizes and shapes, but all work on the same principle: turning the punch's drive screw draws the punch half (below the chassis) into the die half (above the chassis).

Nibbling tool. This inexpensive tool can cut any size or shape hole (bigger than ½-inch diameter) in light-gauge sheet metal. It works by successively nibbling away small bites of metal until the desired hole is formed. You start by cutting a ½-inch pilot hole for the tool's punch, then you operate the nibbler like a stapling gun to chew excess metal away. With care, you can guide the tool with surprising precision. Finally, you use an appropriate file to remove any rough edges.

Files. Start your collection of files with a 10-inch coarse single-cut flat file, a ½-inch-diameter coarse round file, and a small variety of coarse Swiss Pattern files. This is a sufficiently

A tapered reamer is used to slightly enlarge an existing drilled hole in thin-guage metal. Thus it can save you the expense of a comprehensive set of drill bits. Whenever you must make an odd-size hole, use the reamer to enlarge the hole drilled by one of your standard-size drill bits. Also, the tool is useful for deburring the edges of a roughly drilled hole.

comprehensive range for filing aluminum, the metal you will work with most often. You will need a similar assortment of fine-cut files if you also work with steel chassis and cabinets. Two grades of files are necessary because steel and aluminum have radically different physical properties. If you use a coarse file on a hard metal like steel, the teeth quickly dull and chip. And if you use a fine-cut file on a soft metal like aluminum, the teeth quickly clog.

Miscellaneous hand tools. You should also own a metal-working vise (equipped with 4-inch jaws), a hacksaw (with a selection of blades for hard and soft metal), a ball-peen hammer, a selection of C-clamps, and a pair of tin snips or metal-cutting shears. In addition, a set of taps for 6-32 and 8-32 threads is occasionally useful.

Power tools. They are delightful luxuries when you work with electronics, but don't fret if you don't own any. The time and energy they save is probably a lot less than you think, especially if you work with aluminum chassis and cases. For example, drilling a ¼-inch hole through a typical-gauge aluminum chassis with a hand drill takes about 15 seconds whereas a power drill takes about 6 seconds. Since hole-making is the major work involved in preparing a chassis and cabinet for wiring, you can see that the benefits of power tools are limited. Nevertheless, if you own power tools by all means use them.

A *portable power drill* is probably the power tool you will use most often. I have found that my portable drill is far more efficient than a drill press when it comes to making holes in a chassis. This is because most projects require that holes be drilled in the top, and in all four sides, of the chassis. Securing a chassis to a drill-press table without marring it, while the chassis is standing on end, is a tricky and time-consuming job. It's far simpler to clamp the chassis in a swivel-type bench-mounted vise, and attack it from all angles with a portable drill.

Next in line of usefulness is a saber saw. It's a handy tool if you have to cut a large rectangular hole in a chassis. A 24-tooth-per-inch blade is suitable for both aluminum and steel.

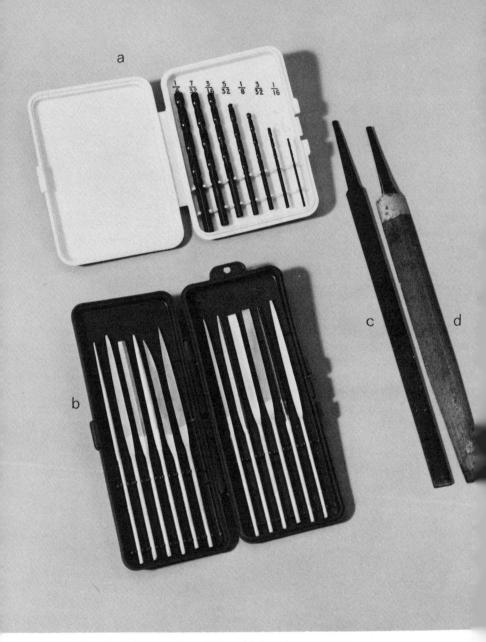

A file of drills and a file of files: (a) the seven drill-bit sizes you should own; (b) a set of Swiss Pattern files; (c) crosscut file; (d) half-round file.

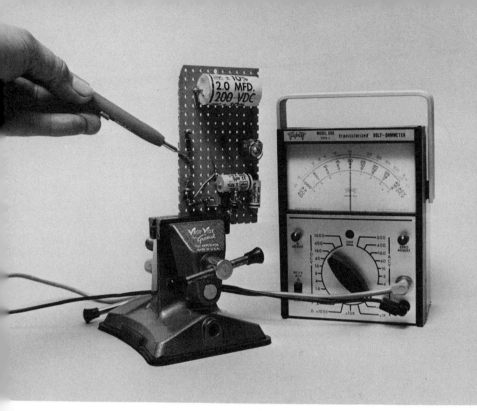

Small modelmaker's vise with vacuum-base becomes a perfect "third hand" to hold circuitry when you troubleshoot or wire.

The only other power tool you will use regularly is a grinder. It's ideal for removing the rough edges from a potentiometer or rotary switch shaft that you've hacksawed down to correct length.

Soldering Tools

For a good solder joint to form, the metals being soldered together, the solder being used to join them, and the flux core inside the solder must all be heated to a temperature of between 525 and 575 degrees F. This is the function of soldering tools. At this high temperature, molten solder flows freely, and is capable of dissolving the surface of any solderable metal very rapidly. (For a complete explanation of the soldering process, see Chapter 6.) I emphasize "solderable metal" since there are many common metals—including aluminum and chromium—that cannot be soldered using the simple techniques discussed in this book. This fact complicates electronic wiring a bit, especially when you are working with an aluminum chassis. All ground connections to the chassis must be made to copper or plated-steel lugs bolted or riveted to the chassis.

The simplest, and probably most useful, soldering tool is the electrically heated *soldering iron*. In its most basic form, it consists of a resistance wire heating element coupled to a *soldering tip*, all mounted at the end of a suitable handle. Most soldering irons are built for 117-volt AC operation, although a few models—primarily miniature high-precision irons—are powered by a low-voltage transformer. Low operating voltage drastically simplifies the required electrical insulation in the handle and heating element, so these irons can be made very compact and lightweight.

Another popular soldering tool, the electric *soldering gun*, works on a slightly different principle. Here, the heating element and the soldering tip are one and the same. A low-voltage transformer in the gun's body feeds a large AC current through the hairpin-shaped tip, to heat it directly. Because the tip reaches working temperature a few seconds after the current begins to flow, soldering guns have pushbutton trigger switches to turn them on when you want to use them. Soldering irons, by contrast, are usually kept at working temperature continuously since their tips typically take upwards of a minute or two to heat up.

When you go shopping, you'll find that there is more to selecting a soldering tool than considering just its style and shape. You must also decide on the tool's electrical power rating, the size of its soldering tip, and, often, the material its tip is made of. The variety is considerable, since there are many different kinds of soldering jobs in electronic wiring, and no one soldering tool is ideal for all.

The power rating of a soldering tool, expressed in watts, specifies how much electrical power the tool draws when it is functioning. Since all of the power fed into the tool is converted into heat, the power rating is a measure of how much

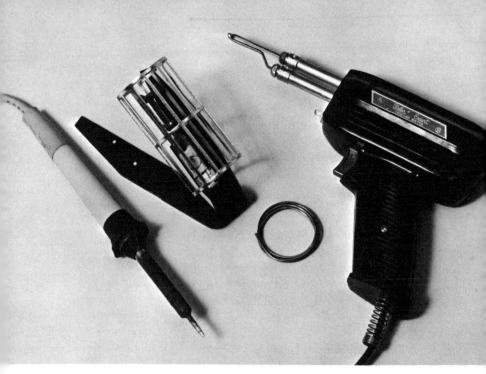

These soldering tools let you tackle virtually every electronic-circuit soldering job. Left, a low-to-medium-wattage pencil soldering iron, with stand, for soldering to printed circuit boards, terminal strip lugs, switch contacts, and most chassis wiring points. Right, a medium-to-high-wattage soldering gun for heavy-duty soldering, such as joining wire to chassis-mounted points.

heat the tool can supply. However, power rating is not necessarily a guide to the tool's working temperature. All soldering tools are designed to reach tip temperatures of between 600 and 1000 degrees F when they aren't actually soldering a joint. Power rating does, however, indicate the *size* of the joint a tool is capable of soldering.

A large solder joint—several wires running to a single terminal, or an electrolytic capacitor lug pressed against a large metal chassis—represents a relatively large mass of metal that must be raised to soldering temperature. This requires that the soldering tool supply a large amount of heat. In short, the tool must have a high power rating (about 75 to 100 watts).

A small solder joint, on the other hand—one wire running to

a terminal, or the leads of a transistor pressed against the foil of a printed circuit board—is a much smaller mass of metal. A soldering tool with a relatively low power rating (about 25 to 40 watts) can supply enough heat to do the job. Often, in fact, you must use a low-power soldering iron for small joints. The transistor in the printed circuit board is a good example. Here, the large amount of heat supplied by a high-power tool could damage the transistor or cause the foil to peel away from the board.

Because of this high-power/low-power conflict, I suggest that you own two different soldering tools: a low-power, pencil-type soldering iron, and a high-power soldering iron or gun. I prefer a soldering gun because of its instant-heat feature. Nearly 90 percent of the soldering you will do when you assemble a kit or build a project requires a low-power iron. It seems silly to keep a high-power iron hot—with the resulting wear and tear on its tip and your electric bill—in anticipation of an occasional large joint. Also, high-power soldering irons are fairly complicated devices. To prevent the enormous heat produced from overheating—and damaging—the tip, most high-power irons are equipped with thermostatic devices to regulate their input power.

What about tip size? Actually, you make that decision when you select a power rating. Large tips go along with high-power heating elements; small tips with low-power ratings. There are many different tip shapes available, too. Here I can offer no advice. You must experiment to find the shape you like best. Most people begin with the simple chisel-shaped tip before trying more exotic shapes.

The material in a soldering tip must meet two requirements: it must be a good conductor of heat (so it transfers heat from the heating element to the solder joint quickly); and it must be capable of being "tinned" easily. Tinning is a simple procedure that coats the soldering tip with a thin layer of molten solder. The idea is to remove the layer of oxides covering the tip so that heat can flow easily from the tip to the joint. In effect, the thin solder layer acts as a heat bridge between the tip and the flux-cleaned metal to be soldered.

Copper meets both requirements superbly, so it is the most common tip material. Unfortunately, copper tips corrode relatively quickly in use. What happens is that as you tin a copper tip, a small amount of the tip dissolves into the thin solder layer. This poses a problem because the tinned surface gradually oxidizes, and must be removed. Thus, each time you brush away the oxidized tinned surface, and re-tin the tip, you are removing a bit of copper. Eventually the tip becomes pitted, and must be filed down to expose clean copper, or replaced.

Iron can be tinned, and it dissolves to a much smaller extent in molten solder than copper. However, it also has much lower heat conductivity. This means that an iron soldering tip takes more time to heat a joint to soldering temperature. Thus, pure iron tips are rarely used. However, iron-covered copper tips are quite popular. They sacrifice a bit of heat conductivity to achieve longer tip life. In addition, several other tip formulas have been developed around the same basic idea: a copper core surrounded by a protective metal sheath or plating.

What tip is best for you? Choose an iron-covered or plated tip for your small pencil soldering iron since it will be hot continuously; you won't have to "dress" the tip as often to expose a clean copper surface. Your large soldering iron or gun can use a less expensive pure copper tip since it will be hot only occasionally, and tip wear won't be a serious problem.

Soldering aids. These low-cost items help you keep your soldering tools working efficiently. A soldering iron stand gives you a place to put a hot soldering iron when you are performing a nonsoldering task on a chassis, and at the same time, lets you keep it in easy reach. A wire brush enables you to scrape off stubborn bits of scale and oxidized material from the soldering tip. A bit of steel wool (or a damp cloth or sponge) lets you occasionally wipe the hot tip free of the oxidized tinning layer before you re-tin the tip.

Desoldering tools. These relatively new devices are primarily designed for industrial use. Basically they consist of conventional pencil-style soldering irons equipped with a set

of special interchangeable tips that are shaped to make removing soldered-in-place components an easy task. A few
models include suction-bulb attachments that suck up the
molten solder in a joint after the tip has melted it.

Electronic Components

Electronic components are the building blocks of electronic circuitry. Designing a circuit really means planning an arrangement of suitable components that will be capable of doing a specific job. And building an electronic device means wiring together a group of selected components according to a specific plan, or circuit design.

Every electronic component has a purpose: to do something or other to an electric current. And just what is an electric current? We now think that it is a free movement of electrons in a medium in which the electrons are loosely "bound" to the atom's core, or nucleus. Electrons have a negative charge, the nucleus a positive charge. Some atoms, say of copper, lose electrons easily; consequently there is a helter-skelter movement of electrons in a piece of wire. Now, if you connect this wire to the positive and negative poles of a battery or to a source of alternating current, there will be a steady movement of electrons (with negative charge) from the negative pole to the positive pole. This flow of electrons is an electric current, which the electronic components in a circuit, in effect, work with.

When you wire an electronic circuit, you must know two things about each of the different components on your work table:

1. How to identify the component.
2. How to handle it when you mount it and solder its leads in place.

Let me emphasize early in this chapter how important this knowledge is. After soldering errors, the two major errors committed by beginners are wiring the wrong component into a circuit location, and damaging components by handling them incorrectly.

Resistors

Resistors are the most common circuit elements and are used to establish correct current and voltage levels throughout a circuit. They can take many shapes and forms, but the resistors you will work with most often are small cylindrical or box-shaped devices, equipped with two metal leads.

Fundamentally, a resistor is a bundle of electrical resistance packaged in an easy-to-handle form. Its heart is a *resistance element*, which can be made from several different substances, and is usually wrapped inside a protective, insulating shell.

Carbon (or composition) resistors are made of small, dowel-like pieces of carbon-based material enclosed in a plastic (or, occasionally, ceramic) case. Different resistance values are obtained by varying the element's length, composition, and diameter.

Wire-wound resistors consist of a length of metal resistance wire wound on a cylindrical ceramic or glass core. Resistance value depends on the length, gauge, and material of the piece

Typical resistors used in electronic circuitry. Left to right: "hum-adjust" potentiometer; standard carbon pontentiometer; 2-watt carbon resistor; 5-watt power resistor; ½-watt carbon resistor (most commonly used rating); adjustable carbon resistor.

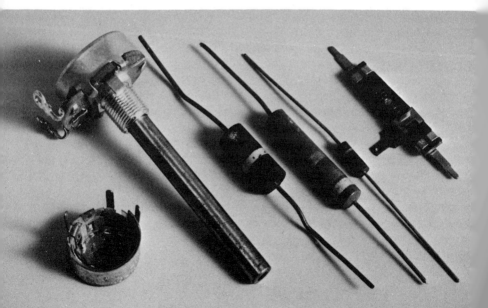

of wire used to make the element. The outer shell is usually made of either ceramic or vitreous enamel (a baked-on mixture of enamel and powdered glass).

Metal film resistors are made by depositing a layer of conductive metal on a cylindrical glass or ceramic substrate (a smooth-surfaced object that will accept the metal coating). The film's thickness, length, and formulation controls the final resistance value. The protective case is most often made from molded plastic or epoxy.

Every resistor, regardless of type, has three important numbers associated with it: its *resistance value*; its *tolerance*; and its *power rating*.

The resistance value, as you would expect, is the resistor's approximate electrical resistance, measured in ohms. How approximate? That depends on the unit's tolerance rating.

Because it is impossible to mass produce electronic components to absolutely exact values, resistors—and other parts, too—are graded according to tolerance—that is, according to how close their actual characteristics match their rated values. For example, the resistors used in measuring circuits usually have a 1 percent tolerance. Their actual resistance values are within 1 percent (they may be higher or lower) of the nominal—or rated—values. This degree of accuracy is rarely needed, and you will usually work with 5 percent or 10 percent tolerance resistors.

The power rating specifies the maximum rate—in watts—at which a resistor can safely dissipate electrical energy as heat, and this point deserves an explanation.

One way to explain how a resistor works is to say that it is a device that converts electrical energy into heat. If this seems odd, think of your car's brakes. They convert mechanical energy (motion) into heat and in the process slow your car down. In a somewhat analogous way, resistors establish desired current and voltage levels not by "slowing down" a current but by "thinning" it, somewhat as a water faucet "thins" water flow or stops it altogether. You see, in a resistor the molecular structure is such that there are frequent collisions between the free electrons and the individual atoms.

These collisions produce heat in the material and the energy of this heat comes from the voltage which is trying to drive the electric current, thus effectively reducing both current and voltage.

Identifying resistors. Depending upon its type and construction, a resistor will be "labeled" with its important characteristics in one of two ways: directly or with a set of color-coded symbols. There are no hard and fast rules, but the trend today is as follows: High-precision and/or high-power-rating resistors usually have their resistance values, tolerance rating, and power rating printed on their bodies. Moderate-precision, low-power-rating resistors—including virtually all carbon resistors, the type you will handle most often—are identified by a group of three or four colored bands painted on their bodies.

Color coding may seem like an unnecessary complication; at first, you will find that it takes a bit of concentration to unravel a banded resistor's identity. The reasons for color coding, though, are good ones. First, the colored bands on a small-bodied resistor are easier to read than tiny printing. Second, and more important, the bands run completely around the resistor's body, making it easy to determine its value no matter in what position it is mounted.

The group of color bands is positioned near one end of the resistor's body to tell you which band to read first. As the illustration shows, the first two bands represent some number from 1 to 99, and the third band represents some multiplier (actually, some multiple of 10, from 0 to 100,000,000). To find the resistor's resistance value, you simply multiply this number by the multiplier. The accompanying table matches up the ten possible band colors with their corresponding digit and multiplier coded equivalents.

The fourth band represents the resistor's tolerance. A gold band means 5 percent tolerance, a silver band means 10 percent, and no band at all means 20 percent tolerance. The power rating of a color-coded resistor is indicated by its overall size.

Variable resistors (called potentiometers or rheostats) are designed into electronic circuitry so that the user can adjust

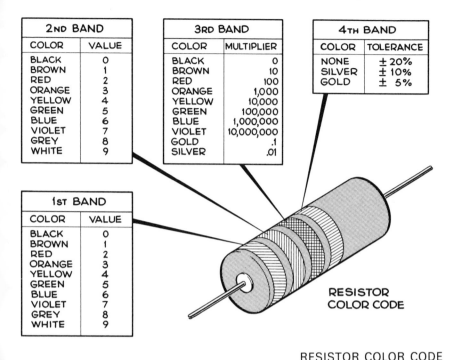

2ND BAND	
COLOR	VALUE
BLACK	0
BROWN	1
RED	2
ORANGE	3
YELLOW	4
GREEN	5
BLUE	6
VIOLET	7
GREY	8
WHITE	9

3RD BAND	
COLOR	MULTIPLIER
BLACK	0
BROWN	10
RED	100
ORANGE	1,000
YELLOW	10,000
GREEN	100,000
BLUE	1,000,000
VIOLET	10,000,000
GOLD	.1
SILVER	.01

4TH BAND	
COLOR	TOLERANCE
NONE	± 20%
SILVER	± 10%
GOLD	± 5%

1ST BAND	
COLOR	VALUE
BLACK	0
BROWN	1
RED	2
ORANGE	3
YELLOW	4
GREEN	5
BLUE	6
VIOLET	7
GREY	8
WHITE	9

RESISTOR
COLOR CODE

RESISTOR COLOR CODE

circuit operation with the twist of a knob. The volume and screen-brightness controls on your TV set are variable resistors; so are the bass and treble controls on your hi-fi amplifier.

A potentiometer—or "pot"—consists of a resistance element (most often made of carbon or wound wire) bent into a three-quarter circle and equipped with a moveable "wiper" arm that slides along the element's surface. The wiper is attached to a turnable metal shaft. By rotating this shaft, the user can shift the wiper to make an electrical contact anywhere on the resistance element, thus putting more or less resistance in the circuit by lengthening or shortening the resistance element.

The three characteristics of fixed resistors—resistance value, tolerance, and power rating—also apply to the resistance element inside a potentiometer. Usually, though, only the resistance value is stamped into the pot's outer protective metal housing. To learn the other two, you must check the appropriate listing in an electronic parts catalog.

Working with resistors. According to one old saw, the only way to damage a resistor is to really try hard. It's true that these devices are so simple that there is little to go wrong, but resistors do have a few weak points. Here are the rules to obey when you work with them:

1. *Avoid overheating.* This is particularly important when you solder the leads of a high-precision and/or low-power resistor. Excessive heat can affect the resistance element's characteristics, and change the resistance value.

2. *Treat the leads gently.* When you mount a fixed resistor between two terminal points, don't place it in a two-way stretch. Too much strain can break the leads loose from the resistance element.

3. *Protect the insulating cover.* Dropping a sharp-edged tool onto a resistor can chip away part of the protective case, exposing the resistance element. The dangers here are that the element will touch the chassis, or a nearby metal object, when the part is mounted, causing a short circuit, and that moisture will reach the element and cause deterioration (an especially serious problem with metal-film resistance elements).

4. *Don't overtighten a pot's mounting nut.* The threaded bushing used to mount a potentiometer can be deformed if you twist the mounting nut excessively, causing the control shaft to bind.

5. *Solder a pot's terminal lugs very carefully.* Use as little solder as possible. Excess solder can creep up the lugs into the housing and ruin the resistance element by short-circuiting it.

Capacitors

Capacitors are devices that possess electrical capacitance. This means that a capacitor is a device that can store—or hold—an electric charge. Large capacitors—capable of storing large quantities of charge—are used as "electron reservoirs" in power-supply stages of electronic circuits. Like water reservoirs, they maintain a constant "pipeline pressure" (which we call the circuit operating voltage) in spite of continuously changing current demands throughout the circuit.

Smaller capacitors are used to "couple" together the

various stages of a multi-stage circuit, thanks to their talent of acting like a short circuit for alternating currents and at the same time like an open circuit to direct currents. And tuned circuits, made up of combinations of capacitors and inductors, are the hearts of every radio and TV set. They have the job of plucking specific signals out of the jumble of different radio waves on the air.

All capacitors contain similar internal structures: a sandwich of two metal conductive surfaces (called *plates*) separated by a thin layer of insulation (called the *dielectric*)—but at this point, most similarity ends. You will work with a variety of capacitors, each of which looks different on the outside, and is built differently on the inside.

Paper capacitors contain long strips of oil-impregnated kraft paper (the dielectric) sandwiched between two strips of thin aluminum foil (the conductive surfaces), and rolled into a tight coil. A wire lead is attached to each foil strip, and the assembly is mounted in a protective cardboard or plastic shell

Assortment of popular capacitor types: (a, b, c) disc capacitors; (d) air dielectric variable capacitor; (e, f, g) plastic-film dielectric capacitors; (h) miniature tubular electrolytic capacitor.

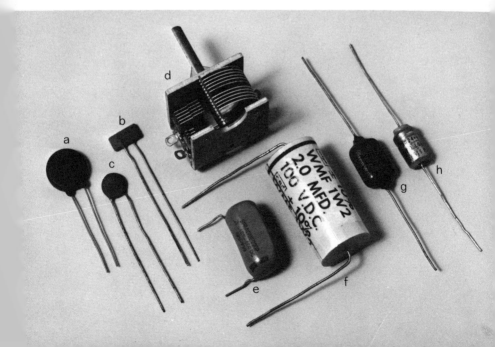

or, occasionally, in a drawn-metal bathtub-shaped case.

Similar capacitors are made by substituting plastic film for the paper dielectric. These are often called "mylar" capacitors, although other types of plastic film can be used.

Mica capacitors use aluminum foil conductors separated by a thin mica flake, and housed in a plastic case. *Silvered-mica* capacitors—a more expensive, high-precision version—have silver film conductive surfaces that are deposited directly on either side of the mica plate.

Ceramic capacitors are available in two forms: the disk ceramic capacitor consists of a thin ceramic wafer wedged between two silver-plated metal plates; the tubular ceramic capacitor is a short length of ceramic tubing whose inner and outer surfaces have been covered with a deposited silver film. Both types are supplied encapsulated in plastic shells.

Electrolytic capacitors consist of two long strips of coiled aluminum foil (as in a paper capacitor) separated by a thin layer of electrolyte—or electricity-carrying liquid. In most modern electrolytic capacitors, the electrolyte is in the form of a thick paste, or is soaked into a strip of thin, porous paper, rather than in true liquid form. At the factory, a DC voltage is applied across the finished capacitor, and an electrochemical reaction takes place on the surface of the foil connected to the positive side of the voltage source. A thin layer of aluminum oxide forms on the foil. This layer—only a few millionths of an inch thick—is the dielectric, and the aluminum strip it covers is one of the capacitor's conductive plates. The other conductive surface is actually the electrolyte (the second aluminum foil strip serves as a terminal that makes electrical contact with the electrolyte).

This unusual way of building a capacitor can pack an enormous amount of capacitance into a relatively small package, but there is a rub: An electrolytic capacitor is a polarized device. Circuits that use electrolytic capacitors are designed to apply a DC voltage across their terminals, in order to prolong the oxide film's life. But the positive side of this voltage must be connected to the oxide-covered aluminum strip. To insure this, every electrolytic capacitor is marked with positive (+)

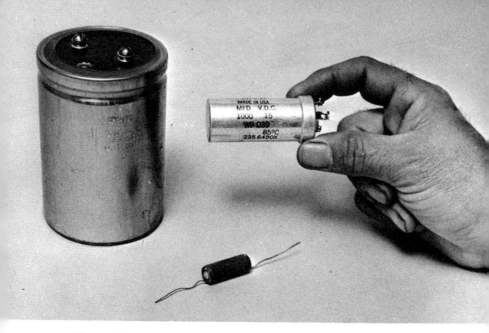

Three electrolytic capacitors: large—a monster unit offering 120,000 microfarads (at 6-volt rating); medium—a familiar "can" type electrolytic designed to mount directly on a chassis; small—a typical tubular electrolytic that is secured by its connecting leads.

and negative (−) symbols next to the corresponding terminals. If an electrolytic capacitor is accidentally hooked up "backwards" in a circuit, the film layer will be destroyed, and the capacitor ruined.

Air-dielectric capacitors. Here, the conductive surfaces are free-standing, thin-gauge metal plates, and air, a good insulator, is the dielectric. The most important member of this capacitor family is the variable capacitor, the device used as the tuning control in the "front end" of most AM and FM tuners and radios. (Car radios are an exception; most use variable inductors instead.)

A variable capacitor consists of a multi-finned aluminum *rotor* that nests within a similarly finned aluminum stator. The rotor is electrically insulated from the stator, and is free to revolve in a pair of bearings. By turning the shaft attached to the rotor, the user can vary the area of the overlapping region between the rotor and stator plates, and hence can vary the capacitance.

Identifying capacitors. The two most important capacitor specifications are capacitance value (measured in farads), and working voltage rating (measured in DC volts). Let's consider working voltage first. This number specifies the maximum allowable DC voltage that can be applied across the capacitor—a higher voltage will break down the dielectric, short-circuiting the conducting surfaces and ruining the capacitor.

The idea of capacitance value seems straightforward enough, but a very confusing complication occasionally arises because the farad is much too large a unit of capacitance to be practical in most electronic design work. The capacitors you will work with will have capacitance values measured in millionths of a farad (or microfarads, abbreviated μF), or millionths of a millionth of a farad (or picofarads, abbreviated pF). *Note:* You occasionally may see the antiquated term "micro-micro-farad" (abbreviated $\mu\mu$F) instead of picofarad.

Confusion can rear its ugly head because there are no rules as to where microfarads begin and picofarads leave off on capacitor labels. For example, a 1000 pF capacitor is also a .001 μF capacitor. The capacitance is the same, but the numbers are different.

Most capacitors are marked directly with their capacitance value and working voltage. The only important exceptions are a few varieties of physically small mica and ceramic units. These are sometimes color-coded, using the same color-to-number code as resistors.

The tolerance of a capacitor is not very significant in most circuit applications, so this information is seldom found on a capacitor's label, unless it is a high-precision unit especially designed for critical circuit use.

One piece of information that's not important to the electronic hobbyist, but is often vital to electronic-circuit designers, is a capacitor's temperature coefficient. This is simply a measure of how much the capacitance value varies with ambient temperature changes, and you may find it listed (usually in a coded or otherwise disguised form) on the capacitors you work with.

Although only electrolytic capacitors are polarized, other

types of capacitors often have one of their two leads identified with a painted band (color doesn't mean anything). The band indicates that the lead is connected to the "outer" foil inside the capacitor (when the foil/dielectric sandwich is rolled up, one of the foil strips ends up on top).

A *multi-section electrolytic capacitor* is, in effect, two, three, or four individual electrolytics packaged together in a single aluminum can. The negative terminals of the different "sections" (each of which may have a different capacitance value) are connected together, and wired internally to the aluminum can, which serves as the "master" negative terminal. The positive leads are wired to a set of terminal lugs mounted on an insulated wafer imbedded in the bottom end of the can. Each lug is identified by a geometrical symbol cut into the wafer, and keyed to a legend embossed or printed on the side of the can.

Handling capacitors. Most of the rules for handling resistors also apply to capacitors (and to all of the components we will talk about in this chapter). Here are a few additional points to keep in mind.

When a capacitor manufacturer specifies a unit's working voltage, he isn't kidding. If you ever must substitute another capacitor for one of a given voltage rating, remember this rule: You can always use a capacitor rated at a higher working voltage, never a lower voltage.

Accidentally reversing the leads of an electrolytic capacitor can be a serious mistake. As we've said, the capacitor will probably be ruined, but, more important, other circuit components may also be damaged. A reversed electrolytic often acts like a short circuit and may cause excessive currents to flow in the other circuit components connected to it.

Don't mix up capacitor types. The different kinds of capacitors we've discussed have different electronic characteristics—even when they have corresponding capacitance and working voltage ratings. When a circuit design calls for a specific type of capacitor, be sure to use it.

Remember that a variable capacitor is a very fragile device. When you handle one always make sure that its plates are fully meshed. It takes very little force to bend a fin and ruin the capacitor.

Inductors

The circuit elements that possess inductance are called inductors, and they are the "flywheels" of electronic circuitry. Their basic physical property is a resistance to changes in the level of current flowing through them, in much the same way that a mechanical flywheel resists changes in its speed of rotation. Thus, physically large inductors, called filter chokes, are often used in power-supply circuits to smooth out the pulses of direct current produced by the rectifier, leaving a continuous DC flow. Smaller chokes are used throughout high-frequency circuitry to isolate the circuit's DC power supply from the high-frequency signals flowing through the circuit. These "r-f chokes" pass direct current readily, but block the flow of rapidly varying high-frequency currents.

Perhaps the most common types of inductors are the members of the huge family of components called "coils." These are the inductors that team up with capacitors to make tuned circuits (we spoke about these earlier) and filters (circuits that act like short circuits for specific frequency signals, and act like open circuits for signals of other frequencies).

The name "coil" pretty much sums up the construction of all inductors: a spool of wire wound on some sort of insulating core (or "coil form"), or around a structure made of iron or some other magnetic material. The variety of inductors is enormous, and the examples that follow are limited to the types you will work with most often:

Iron-core inductors consist of coils of insulated wire (usually enamel coated) wound on cylindrical cores made of thin pieces of steel laminated together. The winding is usually protected by a wrapping of wax-impregnated paper, occasionally by metal covers. Virtually all filter chokes are iron-core inductors.

Ferrite-core inductors are insulated wire coils wound on cardboard, fiber, or plastic tubes that contain a cylindrical slug of ferrite material. (Ferrite is a chemical mixture of sev-

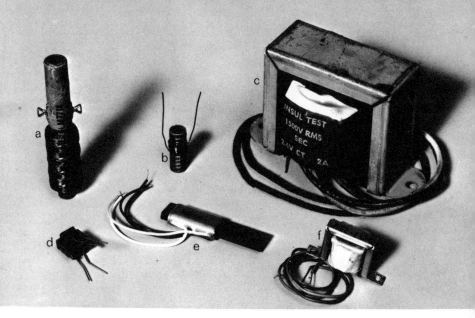

Six coils and transformers you'll probably work with: (a) air-core
r.f. coil; (b) miniature air-core r.f. coil; (c) filament supply transformer (transforms 120 VAC to 12.6 VAC); (d) sub-miniature audio
transformer; (e) ferrite-core r.f. "antenna" coil; (f) miniature audio
transformer.

eral magnetic oxides and an organic binder compressed into
a solid form.) Similar coils can be made using cores of compressed powdered iron.

Adjustable ferrite-core inductors are similar, except that
their ferrite slugs are attached to a thin threaded rod that
rides inside a tapped metal cap mounted on the insulating
tube. Thus, by turning the rod, the user can move the slug up
and down inside the tube, and vary the coil's inductance.

Ferrite-rod inductors. Here, the coil form itself is a rod of
ferrite material. Virtually all antenna coils in modern transistor radios are ferrite-rod inductors.

Toroidal-core inductors. These are a relatively new type in
which the coil is wound on a ferrite toroid—or donut-
shaped—core. For protection, toroidal inductors are often
"potted" (encapsulated) in a plastic shell.

Air-core inductors. This is the name of a very wide assortment of coils that have one thing in common: they are not
wound around any kind of magnetic material. Usually, their

coil forms are cardboard or phenolic tubes, but a few spe-
cies—coils designed to operate at very high frequencies, and
coils designed to work in high-power circuitry which are
wound with thick-gauge wire—use no coil form at all; the
wound wire spiral is self-supporting.

Inductors are electromagnetic circuit elements. A current
flowing through an inductor generates a magnetic field, and it
is the interplay of this magnetic field and the current that
generates it that leads to the device's electrical properties.

Actually, a small magnetic field is produced around any
conductor—even a length of wire—when a current flows
through it. The effect is greatly increased, though, by winding
the wire into a coil, and is further increased by placing a
piece of magnetic material inside the coil (the magnetic material tends to concentrate the magnetic field in a small area,
and aids the interplay between the field and the current producing it).

The inductance of an inductor—measured in henrys—is a
function of its magnetic-field-producing capabilities. Therefore, the greater the number of turns of wire in an inductor,
the greater its physical dimensions; and the greater the amount
of magnetic material inside its core, the greater its measured
inductance will be.

Many of the ferrite- and air-core coils you will work with
will consist of several individual disk-shaped windings, all
connected in series and wound on a single coil form. This
type of construction minimizes "distributed capacitance"
within the coil, an effect that is worth a few words of explanation.

You'll recall that the basic structure of any capacitor is two
conductive surfaces separated by an insulator. At high operating frequencies, adjacent turns of a coil winding act like the
two metal plates of a capacitor; their insulation becomes the
"capacitor's" dielectric. Since each pair of adjacent turns can
in this way act like a tiny capacitor, the total capacitance of
the inductor is distributed throughout the winding.

Distributed capacitance is almost always undesirable, since

it introduces an unwanted—and usually unpredictable—extra component into a circuit, and may affect circuit performance.

Identifying inductors. The unit of inductance is the henry (abbreviated H), but the only inductors that use this unit are physically large devices, such as power-supply filter chokes. The henry is much too large a measure of inductance to specify other types of coils and chokes. The inductance of these devices is measured in millihenrys (thousandths of a henry, abbreviated mH) or microhenrys (millionths of a henry, abbreviated μH). A variable inductor is specified in terms of a maximum and a minimum inductance value—the range of inductance values in between is the device's range of adjustment.

There are four other important numbers that are often—but not always—used to specify inductor characteristics. Whether or not they accompany a particular coil or choke depends mostly on its intended application.

Tolerance (measured in percentage).

Resistance (measured in ohms) of the winding.

Maximum direct-current limit (measured in amperes). This specifies the maximum direct current that can safely flow through the coil or choke. If a circuit exceeds this limit, the inductor will overheat, and may be damaged.

Q—a number that indicates the coil's "degree of perfection." Ideally, an inductor should display no other electrical characteristics other than inductance. Practically, though, an inductor consists of a coil of wire, and this means that all inductors have measurable—although usually quite small—resistances. Fundamentally, the Q of a coil is a measure of how much its resistance will affect its circuit performance. As such, it is of much more importance to circuit designers than to electronic hobbyists.

Unhappily, inductors are the least identified of all common circuit components. Many of the coils and chokes you will work with will have nothing—not even a part number or manufacturer's name—printed or color-coded on their bodies. The obvious rule to live by in this case is to not to separate the component from its wrapping or package (which is normally labeled) until you are ready to solder it in place on a chassis.

The devices that are labeled, usually only have their inductance value (and possibly the manufacturer's part number) printed on the side; to learn the other specifications (you rarely need them) you must turn to a parts catalog.

Color-coding of coils and chokes is rare; the few devices that use colored identification markings follow the same color-to-number key as do resistors.

Tapped inductors. Coils equipped with connections to the interior of their winding, instead of to either end of the winding, are called tapped inductors. Although a tapped coil can have as many taps as there are turns in its winding, the most common types have one, or possibly two, taps. Identifying the tap connection is usually very simple. Its lead or terminal can be differentiated from the end-of-winding connections by studying the winding and/or the layout of the terminals.

Working with inductors. Except for physically large and hefty filter chokes, inductors are among the most delicate circuit components. The reason can be summed up in a single sentence: Most coils and chokes are wound with extremely fine-gauge wire. And therein lies several weaknesses:

The wire, which typically can be 5/1000 inch or less in diameter, is very fragile, and in most coils is protected solely by a thin coat of varnish. A careless slip with a screwdriver can sever the winding or, equally fatal, can break through the insulation of several adjacent turns, and short-circuit them together. Because the wire is so thin, it cannot dissipate heat quickly, and so, when you solder to a coil's leads or terminals, local hot spots can develop which will melt insulation. Therefore, be sure to solder quickly and carefully.

The wide variety of inductor sizes and shapes leads to an equally wide assortment of mounting methods. Heavy iron-core chokes are normally equipped with mounting lugs; physically small ferrite- and air-core coils mount via their leads, like resistors; moderate-size coils (and most adjustable coils) have springy mounting clips formed as part of their metal end-caps.

A few words about this last method: It often takes a substantial push to seat the mounting ears on either side of the chassis or mounting bracket. Be sure you apply this pressure straight down, along the coil form. It is all too easy to crack the form, or dislodge the winding, by forcing the coil home at an angle.

Transformers

Transformers are the "translators" of electronic circuitry. They convert voltages and currents produced in one part of a circuit into voltages and currents required in another stage. For example, the output transformer in an audio amplifier transforms the high-voltage/low-current audio signals developed by the output stage into low-voltage/high-current signals that can drive a loudspeaker.

The range of transformer types used in electronics is enormous—all the way from massive power transformers which convert the power line's 120 volts AC into higher or lower voltages required by power-supply circuitry to tiny intermediate-frequency—or I.F.—transformers which couple the different I.F. amplifier stages together in a radio. (Here, the transformer converts the output signal of each stage into a form that is acceptable to the input of the succeeding stage.)

From the point of view of definition, transformers are inductors. However, practically speaking, their appearance is often different, and the rules for working with them are not always the same, so we've given them a separate section in this chapter.

Fundamentally, a transformer is an inductor that has two coil windings, both wound on a single core. This means that the windings, which are usually electrically isolated from each other, are coupled together electromagnetically. (A few types of transformers have more than two windings—sometimes as many as eight or ten—all on the same core.)

Electromagnetic coupling sounds mysterious, but really is very simple to understand. Suppose an alternating current is fed into winding A of a typical dual-coil transformer. This current produces an alternating magnetic field in the core (the field varies in step with the instantaneous value of current in the coil) which in turn induces an alternating current in winding B.

In a theoretically perfect transformer, all of the electrical energy fed into winding A can be reclaimed out of winding B. In practical units, though, some of the energy is lost for a variety of reasons, so that less energy comes out than went in.

In transformer terminology, the input winding of a transformer (winding A, in our example) is called the primary winding; the output winding (corresponding to winding B) is called the secondary winding. (Transformers with more than two windings usually have one primary and several secondary windings, although there are special-purpose units equipped with multiple primary windings.)

Note that the electrical energy coupled between the primary and secondary of a transformer can be fed in and taken out in different forms. A high-voltage/low-current electrical signal can have as much electrical energy as a low-voltage/high-current signal. This is the key to understanding how a transformer works. By varying the number of turns of wire in the primary and secondary windings, the transformer can be made to accept input signals at one voltage and deliver output signals at another. The ratio of the primary voltage to the secondary voltage is equal to the ratio of the number of primary turns to the ratio of secondary turns.

An important point to remember is that transformers can only work with varying signals, either periodic alternating voltages and currents, or randomly changing signals. There is no such thing as a "DC transformer," and the reason is simple: It takes a varying magnetic field to induce an output voltage in the secondary winding, and only a varying primary signal can produce such a field.

Identifying transformers. All the confusion involved in identifying inductors is there when you work with transformers. Although most power and audio transformers you buy from a parts house will be labeled fully, miniature transformers for transistorized circuits and most I.F. and special-purpose units will probably bear only part numbers and the manufacturer's name.

Generally, the primary and secondary windings of most transformers are terminated at opposite sides of the units. I.F. and other types of high-frequency transformers usually have position-keyed terminal pins so that you can identify the primary and secondary windings; power and audio units *may* have color-coded leads. Transformer color-coding is a widely accepted way of identifying the various leads. However, there is no absolutely agreed upon color-to-lead key, comparable to a resistor color code.

Working with transformers. Everything we've said earlier about handling small components and inductors applies when you work with I.F. and high-frequency transformers. You probably will be handling audio and power transformers most often, and so these units warrant special attention here. In spite of their weight, size, and seeming invulnerability, these types (usually of iron-core design, but occasionally made with toroid cores) have one important weakness: their leads.

The leads you see outside a transformer aren't extensions of the wire used to make the windings; they are soldered on after the core is wound. Thus, an excessive tug on the leads can break them loose, and overheating them can melt the solder joint. And, as an extra hazard, transformer lead wire is about the hardest-to-strip insulated wire you will find. The wax-impregnated insulation seems to fight most wire strippers. So when you strip transformer leads, work carefully. It's very easy to cut through the wire as you remove the insulation.

Vacuum Tubes

Despite a great many claims to the contrary, vacuum tubes are far from obsolete. Although it is certainly true that solid-state devices—including transistors and silicon-controlled rectifiers—have usurped many of the traditional vacuum-tube haunts in circuit design, the tube is still holding its own in many areas. A major reason is cost. In a substantial number of applications it costs more money to build a solid-state circuit that will perform as well as an equivalent tube-type circuit.

There are hundreds of different vacuum tubes available,

but, with a bit of oversimplification, we can classify all of them into two broad categories:

Tubes that rectify.

Tubes that amplify.

(Note that in this section we are not considering TV picture tubes and other types of cathode-ray tubes, "magic eye," or other indicating tubes, and several other kinds of special-purpose tubes that you will seldom work with. However, much of what we say later about identifying and handling vacuum tubes will apply to these devices.)

We will discuss rectification in Chapter 8. Simply, it is the process of converting an alternating current into a direct current; the very first vacuum tube ever built was a vacuum rectifier. Tubes that are designed especially to rectify are called vacuum diodes.

The three elements of a diode are a metal *plate* structure (or *anode*), a *cathode* structure and a *filament* or *heater*, which are all surrounded by an evacuated (vacuum-"filled") glass envelope. In operation, low-voltage AC or DC current flows through the heater, raising it to a high temperature, much like the filament in a light bulb.

The hot filament heats the cathode, which is covered with a metallic oxide that has the ability to emit electrons—quite literally to "boil" them away from its surface—when heated. As a result, a cloud of electrons fills the empty space above and around the hot cathode.

Now, if the anode is made *positive* with respect to the cathode, the electrons will be attracted to it, and a current will flow through the tube. But if the anode is made *negative* with respect to the cathode, the electrons will be repelled away from the plate, and no current will flow.

Thus, a vacuum diode acts like a one-way valve for electric current. And if a diode is connected in series with a source of alternating current, whose polarity changes periodically, current can fiow through the tube during only one half of each cycle, the half when the anode is made positive with respect to the cathode.

The addition of another element called a grid turns a diode

into a tube that amplifies. The grid is a meshlike structure placed between the cathode and plate. Tubes that have one are called triodes.

A simple amplifier circuit that uses a triode is powered by a DC power supply that is connected to the tube in such a way that the anode is always held at a positive voltage with respect to the cathode, so that current can always flow through the tube (and in turn through the resistor connected in series with the tube).

Note that electrons must pass through the grid on their trip from the cathode to the anode—this is possible since the grid is an open structure. However, if the grid is made slightly negative with respect to the cathode, the electric field surrounding it will repel some electrons back to the cathode, and will lower the current flowing through the tube. A sufficiently negative grid voltage will actually "cut off" all current flow by repelling all the electrons back toward the cathode.

Much more interesting, though, is what happens when an alternating voltage is applied to the grid. Now the grid is alternately made more and less negative, and this produces a corresponding alternating-current flow through the tube, which in turn produces a corresponding alternating, but much larger, voltage measured across the resistor in series with the tube.

This is the essence of triode amplification: A small alternating voltage applied to the grid causes a larger—or amplified—replica of itself to appear at the tube's output.

The first successful triode—Lee De Forest's "audion"—was built in 1907. Since that time other amplifying tubes, which include more elements between the cathode and anode, have been developed. The tetrode has two grids; the pentode has three; and the pentagrid tube has five. Although each has specific characteristics that make it the designer's choice for specific circuit applications, all are based on the same basic principle: Voltages applied to the grids control the movement of electrons from cathode to anode.

Rectification and amplification are themselves important

circuit functions; however, they are not the only jobs vacuum tubes can perform. A vacuum diode, in a proper circuit, can become a demodulator (the circuit stage that removes the radio-frequency carrier wave from a radio signal, leaving the sound or music that was broadcast). And amplifying tubes can, with the addition of proper external components, become the hearts of oscillator circuits which generate alternating waveform signals at any desired frequency, or wave-shaping circuits which take an alternating waveform of some particular shape, and change it into another shape. The list of tube applications is almost endless.

Identifying vacuum tubes. There's only one way: by the numbers printed on the tubes' envelopes. From the outside, all vacuum tubes are fairly inscrutable devices—you can't learn much about their characteristics by looking at them. Armed with a number, though, you can consult the pages of a comprehensive tube manual (published by various tube manufacturers) and find out all that is important about the tube that wears it.

By "all that is important" we mean the tube's many specifications, such as: what type of tube is it; which of its internal elements are connected to the several terminal pins on its bottom; what are the maximum voltages that can be applied to its various elements; and what are the tube's intended applications.

The tube number, itself, gives you a very important specification: the correct filament voltage. As in the case of light bulbs, the tube's heating element can be designed to work on almost any voltage. But, once so designed, the tube can not be connected to other values of filament voltage. A lower voltage will not heat the filament to a high enough temperature to insure proper cathode emission; a higher voltage will overheat the filament, and severely shorten its life.

A typical tube number consists of a group of one or two numerals, followed by a group of one or two letters, which is in turn followed by a single numeral. The first numeral group specifies the tube's filament voltage; the rest of the number identifies the tube's specific type. The filament voltage portion of a tube number is always a whole number, such as 3 or 6 or 12. These are really approximations since tube filaments are designed to work on fractional-value voltages—3.15 or 6.3 or 12.6 volts, respectively. Don't try to fathom the identification portion of a tube number; it has no significance other than as an arbitrarily assigned identification code.

A word about foreign-made tubes. The numbering schemes used by foreign manufacturers are considerably different. However, the idea is the same: the code number identifies a specific type of tube. Conversion tables are available that match foreign tube numbers with their American equivalents.

Handling vacuum tubes. Keep in mind that a vacuum tube is an inherently fragile device. Its two weak spots are its glass envelope, and the array of connecting pins on its base.

It's possible—though unlikely—to squeeze the envelope too hard as you remove a tube from its socket, and break the glass. The most common cause of breakage—as you'd expect—is dropping a tube on a hard surface. A word of warning about this: Because the tube envelope is evacuated ("filled" with vacuum), atmospheric air pressure exerts a considerable push (about 15 pounds per square inch) on the glass surface. Thus, if a crack develops in the glass—say as a result of a drop—the pressure can cause an implosion—an almost explosive collapse of the envelope, which can hurl glass fragments a considerable distance. Although this type of hazard is much more pronounced when a large TV picture tube is involved, a conventional vacuum tube can "go off" like a miniature hand grenade if improperly handled.

Up till now we have said very little about the several different sizes that vacuum tubes come in. Regardless of size, their operating principles are identical; size differences are caused by varying internal complexity and the differing physical size of internal elements required to control various electric current levels.

The so-called "miniature tubes" deserve a few extra words, though, because of their unique weakness: the small glass "nipple" that protrudes from the tops of their envelopes. This is actually the remains of a thin glass tube that was part of the envelope while the tube was assembled. It is through this tube that the envelope is evacuated, during the final stages of manufacture; the tube is eventually sealed with a flame, and the excess discarded. Keep in mind that it is all too easy to snap off this little appendage and ruin the tube by destroying the internal vacuum.

As tubes decrease in size, and increase in internal complexity, their base pins become smaller, thinner, and more numerous. Miniature tubes may have 7, 9, or even 11 separate base pins; the new "compactron" tubes (which incorporate three independent tube structures within one envelope) have 13 pins.

As a rule of thumb, remember that it is much easier to bend tube pins than to straighten them out again (even if you own one of the "tube pin straightener" gadgets described in Chapter 3). Try to hold tubes vertically when you insert or remove them from their sockets, and don't exert pressure when you insert a tube unless you are sure that the pins are lined up properly with the socket holes.

"Lined up properly" means that the tube pins are positioned to enter the socket. Every vacuum tube is mechanically keyed, in some way, to insure that it can be plugged into its socket in only one way. The obvious reason is to guarantee that each pin ends up connected to its appropriate circuit components, which are soldered to one of the terminals on the base of the socket.

In a "full size" octal (8-pin) tube, the pins are arranged in a full circle on its base, but a plastic key in the center of the circle forces you to insert the tube in only one orientation.

Miniature tubes have no key. Instead, their pins are positioned in an incomplete circle which matches an incomplete circle of holes on their sockets. The "gaps" in the circles are the index marks that force you to insert the tube correctly.

We will discuss tube sockets in a later chapter devoted to circuit wiring techniques. However, because tubes and their sockets are irrevocably intertwined, they deserve a brief explanation here.

A tube socket is a remarkably complicated device that unfortunately looks very simple. The reason it is complicated is that it has a very difficult job to do: to make a secure electrical connection to each tube pin using nothing more than simple metal-to-metal pressure. And it must keep doing this job over long periods of time, without succumbing to corrosion, even though tubes may be repeatedly removed and inserted into it.

All tube sockets have the same basic structure: a collar of insulating material (usually phenolic or plastic) that supports an array of pin-gripping contacts. Each contact leads to a solder terminal underneath the socket.

Mechanically, tube sockets are quite strong, although it's possible to shatter the insulating material by dropping a heavy tool on a socket. By a wide margin, most tube sockets are damaged by improper wiring technique.

Molten solder flows readily, and if you apply too much solder to a pin terminal when wiring a socket, the excess can flow into the gripping area, and ruin the contact. And, occasionally, overheating the pin will soften the insulating material and distort its shape, loosening its grip on the contact assembly.

Semiconductors

Semiconductor devices get their name from the nature of the materials they are made of: semiconducting compounds. These substances, as the term "semiconductor" suggests, are neither good electrical insulators nor good electrical conductors. This middle-of-the-road behavior makes it possible to build components that can control, switch, and route electrical currents and voltages, the same sort of tasks traditionally performed by vacuum tubes and electromagnetic switches (relays).

The list of different devices and their myriad applications is almost endless. To name a very few, there are *transistors* (which amplify and oscillate); *rectifiers* and *diodes* (which rectify AC signals and demodulate); *controlled rectifiers* (which serve as electronic switches).

The huge class of semiconductor components is often la-

Semiconductor devices are packaged in a variety of unusual case styles: (a) low-power diode or rectifier; (b) moderate-current rectifier; (c) integrated circuit; (d) medium-power silicon-controlled rectifier; (e) audio-frequency power transistor; (f and g) small-signal transistors; (h) cadmium sulfide photo-conductive cell.

beled "solid state." This phrase came into use during the early days of transistorized electronic equipment in order to contrast them with semiconductor devices with vacuum tubes. Inside a semiconductor, current always flows through solid material—not through open space as inside a vacuum tube.

How semiconductor devices operate is a particularly involved process to describe because the mechanisms at work are best explained in terms of the language of quantum mechanics. For the purposes of this chapter, though, we can briefly discuss a simple analogy that comes close to illustrating solid-state device principles.

Think of the chunk of semiconductor material inside any solid-state device as a variable resistor, a unique type that is controlled not by turning a shaft or by sliding a contact, but instead is "varied" by applying voltages and currents to special "control contacts" connected to the chunk. In effect, the control voltages and currents change the nature of the semi-conductor material, varying it from a poor conductor to a good conductor.

Curiously, it was this analogy that helped give the transistor its name. The term "transistor" is really a contraction of the two words TRANSfer and resISTOR. Most transistors are three-lead devices. The electrical resistance measured between two of the leads can be controlled by varying the current fed into the third.

Basically, and remember we are talking in terms of the above analogy, the scores of different semiconductor devices are modifications of the same theme: The electrical resistance of a piece of semiconductor material is in some way controlled by applied currents and voltages.

Identifying semiconductor devices is often a two-pronged problem: The first step is finding out what a particular device is and does (this is similar to identifying a tube by its code number); the second—and equally important—is to figure out how to connect the device to the circuit you are working with.

To a large extent, semiconductor devices have code numbers, like vacuum tubes. Unfortunately, there is nowhere near the same degree of standardization among solid-state component manufacturers as there is in the vacuum-tube industry. The sad facts of semiconductor life are—with all too few exceptions—that a particular type of device may be made by only one or two of the scores of different manufacturers. And, consequently, manufacturers are often tempted to tack their own, individually created designations on their products, rather than go to the trouble of having an impersonal "official" number assigned to each one.

Thus, if you flip through the semiconductor listings in a parts catalog, you will find diodes listed that have 1Nxxx code numbers (such as 1N695). The 1N means that the device bearing it has two leads. And you will see hundreds of transistor types carrying 2Nxxx numbers (such as 2N170)—here, the 2N means each transistor has three leads.

But you'll also find hundreds of diodes and transistor types listed that have numbers such as SK-3007 (a transistor) or 8D6 (a rectifier), numbers that were arbitrarily tacked on by the manufacturer.

From a purely practical standpoint, these "unofficial" numbers are just as useful as "official" numbers since, once armed with one, you can look up the device's specifications in a catalog. But there's a rub: You must know the manufacturer as well as the number.

Diodes and transistors aside, there is a legion of other semiconductor devices that are all labeled with manufacturer-assigned numbers. Here there simply are no official numbers yet. A few of these, which you will probably work with often, are: *silicon-controlled rectifiers*; *"triacs"*; *photoconductive cells*; and *integrated circuits*.

In at least one way, semiconductor devices are similar to vacuum tubes: You can't be sure of a particular device's function just by looking at it. The reason is that there are many "standardized" packages—or housings—for semiconductor devices, and often many different types of devices may be housed in similar cases.

Deciding which lead or terminal goes where, when you are wiring a semiconductor device into a circuit, is usually a thorny problem. It's easy to make mistakes, and a wiring error can be fatal to a semiconductor component when you switch on the power to the "finished" circuit.

The cause of this problem is the physically small size of most semiconductor devices. Unlike vacuum tubes, many have thin, flexible leads, not plug-in pins, so mechanical keying systems can't be used. And, anyway, sockets are rarely used these days to mount even those solid-state components that lend themselves to socket mounting. Most are just supported by their leads, which are soldered to terminal points or to a printed circuit board. Slightly larger devices bolt directly to a chassis (or often to a "heat sink") via a threaded stud machined into the housing. These units usually have protruding terminal lugs (often the metal housing is one of the connecting "terminals"). Other devices—including many power transistors (transistors designed to control large currents and voltages)—have mount-

ing holes drilled into their cases. These units are also bolted
to a chassis when they are installed. The case represents one
of the connections; other terminal pins protrude from the
bottom.

As a rule, the various terminal leads or pins on these solid-
state devices are labeled with names, rather than numbers (as
are the pins on a tube). The names used are the technical
names of the different structural elements inside the device.
For example, the three leads of a transistor are labeled
"Emitter," "Base," and "Collector"—meaning that each lead
so named is connected to that particular element inside the
transistor. And the two leads of a diode are labeled "Anode"
and "Cathode."

Most devices, however, are too small to carry the lead designations on their bodies in words or letter abbreviations, and
so other coding schemes are used. Although they seem numerous—a different coding scheme for each type of device,
and usually for each different type of housing—all are based
on the same idea: positional coding that is keyed to some
physical feature of the housing.

This physical feature may be simply a colored dot or band
painted on the housing; it may be a small protruding index
tab; it may be a small flattened area on the housing; or it may
be the unsymmetrical positioning of the terminal leads as they
emerge from the housing. The function of this feature is to
tell you how to orient the device when you look at it. Once
you've done this, you can identify the various terminals or
leads from their relative positions on the housing by consult-
ing a case or base diagram for the device.

Handling semiconductors. A paradox of solid-state electronics is that semiconductor components, which are inherently rugged, can be damaged in many different ways. The
ads for solid-state electronic gear suggest that transistors,
diodes, and the host of other semiconductors used today, are
virtually indestructable. This has proved true in many cases
where the semiconductors were properly installed in well-de-
signed circuits. Since in this book we are more concerned
with the installation end of circuitry than with design, we will
consider this aspect in this section.

Mechanically speaking, all the solid-state components that you will work with have few weak spots. Dropping a transistor or diode probably won't damage it, although there is a slight chance that the shock will crack loose one or more of the internal welds that connect the terminal leads to the chunk of material inside the housing, which will ruin the device. The thin leads used on small components are obviously fragile. The rule is to avoid flexing them unnecessarily, to minimize the risk of metal fatigue, which could cause them to break.

Electrically speaking, though, semiconductors are about the most unforgiving components you will handle. Hooking their leads up incorrectly (and in some cases misconnecting the leads of components wired to them) can destroy them in a very expensive flash. We will cover this point again in a later chapter on wiring techniques, but keep in mind whenever you install solid-state components that a wiring error can be an expensive mistake. Be sure you have identified each component lead correctly, and that you know which lead goes where.

Heat is another potential enemy. Overheating solid-state components when you solder their leads can ruin the delicate semiconductor devices inside the housings. Be sure to follow the soldering rules for these devices, given in a later chapter.

Photocells

Photocells are the "electric eyes" of electronic circuitry, and are being used in increasing numbers in scores of applications, especially photographic equipment and home surveillance devices (photocell door annunciators and burglar alarms, for example).

There are two types of cells that you will work with most often:

1. Photoelectric cells, miniature light-powered electrical generators that convert light energy striking their surfaces into electrical energy. Depending upon the semiconductor material they are made of (selenium and silicon are the most common raw materials), typical cells can produce a current

ranging from between several microamperes and a few milliamperes DC, at a voltage of approximately ½ volt DC, when placed in bright sunlight.

2. Photoconductive cells, which are light-controlled resistors. The electrical resistance of this type of cell is proportional to the intensity of the light striking its surface: the greater the brightness, the lower the cell's resistance.

The operation of both kinds of photocell is complex. Inside a photoelectric cell (which you occasionally may see labeled "photovoltaic cell"), the cell's geometry establishes an electron barrier. Photons (packets of light energy) striking the cell, dislodge electrons and propel them across the barrier, building up an electron concentration on one side and, consequently, developing the voltage produced by the cell.

Photoconductive cells use a totally different mechanism. Here, light striking the cell's surface causes an increase in the number of free—or moveable—electrons. As the number of free electrons increases, the material changes character from an insulator to a conductor. It's electrical resistance decreases.

Identifying and handling photocells. There is no standardized labeling system for photocells, so, once again, you are at the mercy of the manufacturer's code number printed on every cell you work with. The important characteristics of a photoelectric cell are its output current and voltage (measured at some specified illumination level when connected across a specified "load" resistor) and the polarity of its leads or output terminals. The first two characteristics are usually found in a catalog listing; polarity is normally indicated by lead color-coding (red for positive) or a + sign next to the positive terminal. (*Note:* The "polarity" of a photocell is similar to the "polarity" of a battery or other source of direct current; it indicates the direction of generated current flow produced by the cell. In other semiconductor devices, the term "polarity" usually signifies the correct direction of current flow through the device.)

The key characteristics of a photoconductive cell are its light and dark resistance (resistances measured when the cell's surface is in darkness, and at a specified illumination level);

the cell's maximum power dissipation in watts (equivalent to the power specification for a resistor); and, occasionally specified, its maximum voltage capability (the maximum DC voltage that can be applied across the cell safely). Photoconductive cells are, like resistors, bilateral devices; you can install them in either direction, so there is no "polarity" indication.

All of the semiconductor handling precautions apply to photocells in general; photoconductive types deserve extra care because of their glass elements. Depending upon construction, either a cell's front-surface window, or its entire case, will be made of glass.

Switches
Fundamentally, any device that controls or directs the flow of an electric current can be thought of as a switch. This broad definition encompasses a great variety of electronic, electromechanical, and purely mechanical devices. Therefore, in this section, we will limit the discussion to the mechanical types that you probably think of when you hear the word "switch."

The way a switch is named, or labeled, invariably describes its mechanical innards, and is generally a two-part name:

1. Number and kind of switch *contacts*.

2. Geometry or structure of the switch.

We'll consider the second part, first. There are many kinds of switches; the most common are as follows:

1. *Knife switch*—the most ancient switch structure, which dates back to the eighteenth century.

2. *Toggle switch*—to insure a positive switch-open and an equally positive switch-closed state, the mechanism uses a mechanical spring-loaded "seesaw" arrangement. This guarantees that the switch will be either open or closed, and can't be any place in between.

3. *Pushbutton switch*—the spring-loaded shaft doesn't activate the contact assembly until the pushbutton is pressed. Contacts can be designed to open, or to close, when the button is pushed.

4. *Slide switch*—an inexpensive design in which the plastic knob slides the "wiper" assembly to make or break the connection.

5. *Lever switch*—the lever is part of a mechanical linkage that opens or closes the switch contacts when the lever is "thrown."

6. *Rotary switch*—the moveable shaft carries the wiper to any of the several contacts mounted on the insulator's periphery.

Switch contacts are described in terms of *poles* (or number of sets of contacts), and *throws* (number of different switch positions). An exception is rotary switch terminology: the term "throw" is replaced by "position."

The simplest possible switch contact arrangement is the

Typical switches and pilot lights used in electronic projects: (a) lever-action rotary switch; (b and d) pushbutton, momentary contact switches; (c) toggle-action switch; (e) preassembled neon-bulb pilot lamp assembly (includes current limiting resistor wired in series with neon bulb); (f) low-voltage lamp pilot-light assembly.

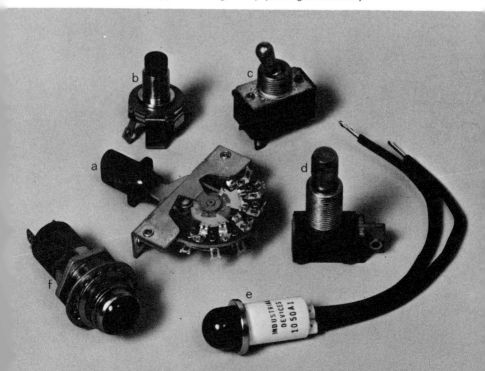

single-pole, single-throw, abbreviated SPST. Here, a single set of contacts closes or opens to control a single current.

A single-pole, double-throw arrangement (abbreviated SPDT) has a single moveable contact that can be "thrown"— or positioned—to either of two fixed contacts. This is the electrical equivalent of a fork in the road.

In some of the switch geometries, the number of positions can be increased to three or four, and a rotary switch can have dozens (literally) of different contacts for the moving contact to touch.

By ganging the above basic contact arrangements, a wide assortment of complex switches can be created. For example, a double-pole, double-throw (DPDT) contact set is simply two independent SPDT contact sets operated by one switch mechanism. And a 12-pole, 24-position rotary switch consists of 12 individual 24-position rotary switch "wafers" mounted on a single shaft.

When you combine a particular contact arrangement with a specific mechanism, you produce—or specify—a unique switch model. For example: A DPDT toggle switch, or a 3-pole single throw (3PST) slide switch, or a SPST lever switch.

One final point: Spring-loaded switches, such as pushbutton types, are available, as we've said, so that the contacts are either opened, or closed, when the switch is operated. A single-pole, single-throw pushbutton switch with normally open contacts (the contacts close when the switch is activated) is labeled SPST-NO (the "NO" is an abbreviation for normally open). And, logically, a SPST-NC switch has normally closed contacts; they open when the switch is operated.

The contacts themselves—the most important element of any switch—are made of a variety of materials ranging from brass to elaborate alloys of precious metals. It all depends on the price of the switch. The most important switch-contact specifications are maximum-rated current and voltage, and (occasionally) typical number of operations—the manufacturer's estimate of how many times the contacts will properly make and break a circuit.

You can generally identify a switch by its physical appearance, and the arrangement of terminals on its body. However, for the important specs, you'll have to check a catalog most times. A few switch manufacturers print voltage and current limits on the bodies of their products.

Switches, as a group, are among the most rugged components you will handle because of their essentially mechanical nature. However, it *is* possible to damage them by:

• Using too much force when you tighten their mounting nuts. Excessive torque can distort their bodies and prevent smooth mechanical operation.

• Using too much solder when you wire their terminals, or by doing a sloppy soldering job. Excessive or carelessly applied solder may creep up the switch terminals and enter the contact mechanism, and ruin the switch. This is an especially serious hazard when you handle rotary switches.

• Using too much force when you secure wires to the terminals. A vigorous pull with a pair of needlenose pliers can snap off the terminals on many switches, especially miniature units.

Relays

Relays are electromechanically operated switches, which is another way of saying that they are switch contacts that are activated by an electromagnet. The idea is simple: In many electronic applications it is desirable to have a switch that operates electrically, that is activated by a flowing electric current. And a relay is just that.

Practical relays come in an enormous assortment of sizes, shapes, and capabilities. Coils can be designed to operate across a wide range of AC or DC currents and voltages, and the contact assembly can be built with a staggering assortment of contact arrangements.

By and large, the same terminology we discussed above, for switches, is used to describe realy contacts; the only additional important specification being the particular voltage and/or current rating of the coil.

Relays designed for industrial use have a standardized labeling system, but this is of little help to the electronic hob-

Relays are electromagnetically actuated switches. Left, a typical low-cost moderately sensitive DPDT-contact relay. Right, an ultrasensitive dual-coil relay equipped with adjustable SPDT contacts.

byist who almost invariably works with nonindustrial types. In most cases, though, the only important specs you may have to look up in a catalog are the maximum current and voltage ratings of the contacts; coil voltage and/or current requirements are usually printed on the coil, and the switch contact arrangement can be determined by inspection.

Hardware

Many items of electronic hardware will be familiar to you already; others will probably seem curious until you actually work with them, especially the hardware used in electronic wiring. Since a large part of the remainder of this book is concerned with the correct and proper use of these components, we will simply catalog them in this section, to provide you with an illustrated reference to hardware items.

Basic hardware. Conventional machine nuts and bolts are used to mount components and assemble chassis and cases. The most often used sizes are 2-56; 4-40; 6-32; and 8-32.

Lockwashers are used between machine nuts and metal chassis; and between small metal components and metal chassis, to prevent bolts and nuts from loosening.

Rubber grommets (and nylon grommets, a new development) are used to insulate the rough edges of holes cut

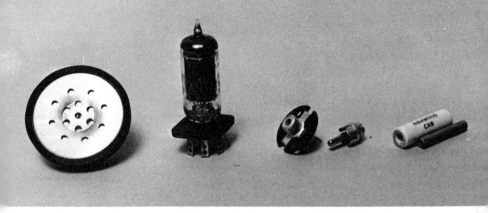

A miscellaneous group of components. Left to right: miniature crystal microphone; miniature tube socket (shown with tube); audio connector and its mating plug; pair of spacers used to hold major components and subchassis above or away from chassis surfaces.

through chassis, so that wires can be passed through them safely.

Insulators and metal spacers are used to support subchassis and other components away from metal chassis.

Shaft hardware is used to mount shaft-equipped components (potentiometers, variable capacitors, and rotary switches). Shaft bearings carry the shafts through instrument panels; universal joints allow shafts to be bent around large chassis components; shaft couplers allow extension shafts to be joined to standard-size shafts.

Chassis are the skeltons of electronic wiring, and are available in a wide assortment of shapes and sizes. Aluminum and thin-gauge steel are the most common materials, although copper or copper-plated steel chassis are used occasionally.

Metal and plastic cabinets, the outerclothes of electronic circuitry, also come in a staggering variety of shape, size, material, and finish.

Terminal strips are used as soldering points for jointing wires together, and as mounting points for small components. A variety of styles are available, all built pretty much on the same idea: a number of metal terminal lugs supported and mounted on an insulating base.

Perforated phenolic board is one of the most popular chassis and subchassis materials among electronic hobbyists. It

A pre-built and wired amplifier module circuit. The device mounts to a chassis via its two control-shaft bushings.

Aluminum miniboxes and chassis are available in a wide variety of shapes and sizes to accommodate almost any imaginable circuit.

looks like miniature Peg-Board, and works in much the same way. You use it to create any circuit arrangement you choose by inserting *push-in terminals* in selected holes that serve as mounting and wiring points for the circuit's components.

Connectors serve to join wires and cables, and are the input and output terminals for electronic circuitry. The types you'll work with most often are audio-frequency and radio-frequency connectors; audio plugs and jacks; binding posts; banana plugs and jacks; pin plugs and jacks; Jones plugs and jacks; barrier terminal strips (really connectors); and clip-on connection devices for ends of wire such as alligator clips.

Solderless terminals are applied to the ends of wire and cable with a suitable crimping tool. They are used extensively in automotive wiring, and are becoming increasingly popular throughout electronic wiring.

Pre-wired modules are factory-wired solid-state circuit stages that perform complete circuit functions but act like simple components. Typical modules include audio amplifiers, oscillators, power supplies, and burglar-alarm circuitry. You handle the module as if it were a "black box" component: Bolt it in place in a cabinet or on a chassis, and hook up the appropriate input, output, and power-supply leads. Occasionally, modules require additional components to work. For example, an amplifier module may need an externally mounted volume control. Many modularized circuits are complete enough to act as the heart of a simple project all by themselves. You'll find a selection of these listed in most electronic supply-house catalogs; they are often supplied complete with simple schematic diagrams that outline typical project circuitry.

Wiring harness. This is used extensively by kit manufacturers and commercial electronics companies. Essentially, a harness is a bundle of color-coded wires, each precut to a specific length, which are laced together. The wires are positioned in the bundle in such a way that when the harness is placed inside a chassis, the ends of the wires are located adjacent to the terminals of components and terminal strips mounted on the chassis. Thus, all of the chassis-mounted parts can be interconnected quickly and neatly—without the use of many individual wire interconnections.

Battery holders are used extensively in small, low-power transistorized devices. They range from simple chassis-mounted clip-type holders (available in many different sizes) to elaborate insulated tubes.

Heat sinks. A major problem of semiconductor circuit design is removing the potentially damaging heat produced by high-power solid-state devices when they operate. The most common solution is to mount the device on a heat sink, a large mass of metal (usually aluminum) that is shaped and finned to dissipate the heat quickly. Heat sinks come in a wide assortment of sizes and shapes to accommodate different semiconductor devices.

Fuses and circuit breakers. These are scaled-down versions of the units that protect your home's wiring. Different holders and mounting devices are available to permit both chassis-mount and panel-mount arrangements.

Pilot lamps and holders. Scores of lamp types and dozens of holders are available, to operate on a wide assortment of voltages.

Fans are another solution to the heat-dissipation problem, although still relatively rare in consumer electronic gear. Electronic chassis fans are designed to move moderate volumes of air without making appreciable noise.

Mechanical Assembly

Mounting the major components—the circuit parts that are mechanically fastened to the chassis or cabinet—is the first actual assembly step you will perform when you build a project. It's an important step, and how good a job you do will affect the project's appearance, its electrical operation, and its mechanical sturdiness.

As you'd expect, mounting each component involves more than simply wielding a screwdriver or a wrench. When you build a project from scratch, you must provide the mounting holes. You may have to prepare various mounting surfaces, and you will have to select the mounting hardware. That's why we have given this step a suitably all-encompassing name—"mechanical assembly"—and are devoting this chapter to the important techniques you should understand.

Incidentally, since the very same mechanical assembly techniques apply when you work with either metal chassis or metal cabinets, we'll save a bit of breath by talking only about "chassis," rather than "chassis and cabinets."

Making a Punching Diagram

Before you can begin drilling holes, you must convert the parts layout plan into a simpler diagram: a map of the necessary mounting holes for the major components. This punching diagram shows the location, shape, and size of every hole and every opening you will cut in the chassis.

Start by selecting the proper mounting hardware for all components that must be bolted to the chassis. A few parts (large variable capacitors, for example) are tapped to receive machine screws directly; the majority, though, have screw holes or mounting lugs that each require a screw-and-nut fastening pair.

Virtually all of the machine screws and nuts you will need to build electronic equipment fall into four size groups: $2/56$; $4/40$; $6/32$, $8/32$. You should own an assortment of screws in these sizes: ¼ inch; ½ inch; 1 inch; 2 inches, along with appropriate nuts. Both screws and nuts should be made of nickel-plated brass; screws should have binding-type heads.

As a rule of thumb, always plan to fit the largest size screw that will pass through a lug or mounting hole; and plan for a screw in every mounting hole or lug provided. Four-corner fastening is better than two-corner support.

Consult a screw-hole size chart for the proper mounting-hole diameter for each type of screw that will be used. Then, draw a carefully dimensioned sketch of the chassis that shows the mounting-hole locations for each bolted-on component.

Indicate the size of the drill bit you will use. Don't specify a bit more than $1/32$ inch larger than the proper diameter. Instead, if your bit collection doesn't include a particular size that is called for, use the closest smaller bit, and then use a tapered reamer to enlarge the hole to its required diameter.

Several familiar components that bolt in place, including panel meters, tube sockets, and some large transformers, require chassis cutouts. These may be large round holes or odd-shaped openings, but their dimensions are usually critical. And so you should make a cardboard or paper template for each component (if none is supplied), and use it to draw an outline of the required cutout on the punching diagram. Accuracy is important here, because the positioning of the cutout must agree with the location of the component's mounting holes. Remember, when the chassis is "punched," each component must fit within its cutout, and its mounting lugs must line up with the chassis holes.

Next, consider the panel-mounting components—parts that pass through the chassis and are held in place by a large nut tightened on a threaded shaft (most switches, many fuseholders, and almost all potentiometers) or by a metal spring clip slid along their body (many neon-bulb pilot light assemblies). Add their mounting-hole locations and diameters to the diagram. A point to keep in mind: The two most common threaded-shaft diameters are ⅜ and ½ inch, and it's difficult to tell them apart visually. Thus, it's a good idea to measure every component before you start drilling.

Insulated components, such as binding posts, banana jacks, and some power transistor sockets, pose a special problem. The size of their mounting holes is determined by the diameter of the insulating collar placed inside the hole. This collar prevents the mounting screw or stud from touching the sides of the hole and short-circuiting the component. Be sure to measure collar width very carefully—the size of the hole you will drill is critical. Too large a hole may allow the collar to slip out of place; too small a hole won't permit the collar to seat properly. Both mistakes could cause a short-circuited part.

Rubber grommets can be considered as panel-mounting components, since they pass through the chassis surface when installed in a hole. I prefer, though, to treat them as part of another group: wire-guiding components. The labeled size of a grommet—in inches—refers to its inner open-area diameter; its mounting hole must be larger to accommodate the thickness of the rubber inner rim. Typically, you can add ⅛ inch to the grommet size to calculate the required mounting-hole diameter; for example, a ⅜-inch grommet mounts in a ½-inch hole.

Variations are possible, though, so I suggest you measure the innver diameters of the grommets you use. Hold the grommet edgewise up to a light so you can see its cross section in silhouette. Compare the inside diameter with the scale on a transparent plastic ruler held just below the grommet.

Cable clamps, another wire-guiding part, can be tricky to deal with, since you are forced to guess the exact path of wires you haven't yet soldered in place. Their function, as their name implies, is to hold connecting wires close to the chassis, and help keep wiring neat. In most cases it is impractical to install a cable clamp until chassis wiring is complete; on the other hand, it is desirable to drill its mounting hole before wiring begins. Happily, this paradox has three solutions:

Rubber grommets are wire-guiding components. They permit wires to pass through chassis openings without the danger of sharp metal edges cutting through insulation.

1. Estimate clamp location by considering the schematic diagram and the parts layout.

2. Plan to mount cable clamps with the same screws holding other components.

3. Plan to drill mounting holes—using a hand-powered drill—after the wiring is completed. Quite clearly, extreme caution is required.

Finally, add the location and sizes of the mounting holes for the various *subchassis*, if any, to the punching diagram. By subchassis, we mean wired perforated chassis boards; aluminum heat sinks; prewired modules; metal brackets holding controls or switches; and the like.

When bolt holes are already drilled in these components, treat them like other bolt-in-place parts; if you must drill holes, use the following ground rules as guide lines:

• $6/_{32}$ machine screws are the best compromise between small size and high support strength, and will be adequate for fastening most subchassis.

• My rule of thumb is to provide four support and/or fastening screws for any typical subchassis measuring up to 4 square inches. I add two more screws for every additional 2 square inches (or fraction thereof).

• Plan to provide additional screws when a subchassis carrying heavy components (such as a large transformer) is supported above the main chassis by insulating spacers placed over the mounting screws. Add two more screws close by each heavy part.

• Edge or corner support is not sufficient for a relatively flexible perforated-board subchassis that carries heavy components. Add additional supporting screws at points along the centerline, using your judgment.

• Add angle brackets, if necessary, to support subchassis mounted close to chassis corners.

• Components mounted on support brackets may require additional support—a heavy part on the end of a long base-fastened bracket may cause the bracket to flex with equipment motion, unless reinforced.

Drilling Holes in the Chassis

You start with a bare metal chassis, a hole-punching diagram, and a toolbox. Your end product is a punched chassis, neatly laced with the proper assortment of mounting holes and cutouts, hopefully all cut to the right size, and placed in the right location. The straightforward ten-step process to achieve this goal is described below:

1. Protect the surface of the chassis before you do anything else. This is really an important help in creating a professional-looking chassis. If the chassis comes wrapped in paper, leave the wrapping on; if not, place strips of broad masking tape over the areas you will be drilling. The paper or tape provides a "writing surface" you can use to indicate hole locations, and it helps prevent drill and/or tool slippage. *Note:* If you will be cutting large round holes (larger than the size of your drill bits or hole punches) or odd-shaped openings, you must use masking tape instead of wrapping paper (unwrap any chassis so supplied). The reason is that paper will tear and come loose during the cutting operations required by these openings.

2. Carefully transfer the locations of the required holes to the paper or tape, with a sharp pencil, and, in the case of outsized holes or odd-shaped openings, use the templates you

made earlier to draw an actual outline of the opening on the masking tape.

3. Centerpunch each X to provide a starting depression for the drill bit.

4. Decide how you will cut large round holes and odd-shaped openings. There are two practical methods:

a. Drill a pattern of small holes around the inside of the opening's intended perimeter, file away the small bits of metal that hold the inner "slug" to the chassis, then file the rough edges smooth.

b. Punch one or two moderately large starting holes (say ¾ inch diameter), use a metal nibbling tool to remove most of the unwanted metal, then smooth rough edges with a file.

As a rule, the first method produces neater large round holes; the second, cleaner square or rectangular openings. Both, though, can be used interchangeably if you work carefully. When you decide, centerpunch the starting points of either the perimeter holes or the large starting holes.

5. Back up aluminum chassis with scrap wooden blocks before you drill. The relatively thin-gauge aluminum will often bend under drill pressure.

6. Drill the "small" round holes (those that won't be enlarged later, or aren't guide holes for chassis punches) carefully, using as little drill pressure as possible. The idea is not to break through the thin layer of metal as the drill bit nears the bottom surface of the material. This causes hard-to-remove jagged edges around the hole. Backing up the metal with wood, as described above, also helps prevent this.

7. Use a tapered reamer to enlarge holes, as required, and to make guide holes for drive bolts. Note that you do not rotate a reamer continuously, like a twist drill or corkscrew. The best technique is to turn the reamer about half a turn clockwise, at the same time applying downward pressure. Then, without releasing the pressure, turn the reamer a quarter turn counterclockwise. This helps sever the metal shavings cut during the first half turn. Alternate these half- and quarter-turn opposite-direction twists until the hole is enlarged to the proper size.

Ream all guide holes about $\frac{1}{16}$ inch wider than the drive-bolt diameter. The reason is that the center slug removed by the punch will wrap around the bolt as the punch works. This little bit of extra clearance makes it easier to remove the slug at the end of the job.

8. Use your chassis punches to cut the required large holes. A chassis punch is a screw-driven machine. When you turn the guide bolt with a wrench, a sharp-edged metal punch is forced through the surface to be cut, into a metal die. The "punch side" of the hole will have smooth, rounded edges, made as the punch squeezed by; the "die side," though, will have a sharper and slightly raised edge. Thus, if chassis shape permits you to swing a wrench inside, set up the chassis punches so that their punch mechanisms will cut inwards, through the outer surface (the visible surface) of the chassis.

One very important point: Always remove the slug after you punch a hole; don't let slugs accumulate inside the chassis punch.

9. Cut the large and odd-shaped holes using one of the methods described above. The fastest way to nibble a hole is first, to cut away a rough-shaped slug about ¼ inch within the perimeter of the desired opening. Then, use the tool to nibble away the remaining unwanted material. Get as close to the outline as you can, but be careful—you can't unnibble the edges. Finish up by filing the edges smooth.

10. Remove the wrapping paper or tape, and examine the holes for rough edges or dangling metal whiskers. Sometimes these can be filed away without damaging the surface of the chassis, but the technique I use most often is to carefully push the rough metal slivers towards the center of the hole with a narrow-blade wood chisel and then slice them off with a reamer or narrow file.

Plastic boxes. Cutting holes in plastic can be as much of an art as it is a science. Plastics crack easily, they soften when worked and clog drill bits and files, and they defy the use of time-savers like chassis punches or nibbling tools.

The two best hole-drilling tools are a hand drill, or a power drill slowed down to ¼ speed (400 to 500 rpm). Use a variable-speed drill or a power-tool speed control. Apply as little drill pressure as possible, and don't drill holes larger than ⅛ inch. Use a tapered reamer to enlarge holes from there.

An alternate technique that works with clear plastic (not bakelite) is to punch a starting hole with a very hot needle or ice pick. Then continue with a reamer.

Preparing the Surface

Surface preparation is required for the mating surfaces beneath screw-fastened components that must make a good ground connection to the chassis, and for the chassis surface beneath insulating washers and/or collars.

To insure an intimate electrical connection between lug and chassis, remove all rough edges from the mounting hole, and scrape away any paint from the surface that will be directly under the lug *and* the screw head.

Insulating washers—especially the thin mica type used to insulate power transistors from chassis and heat sinks—pose a special problem. The smallest sharp-edged metal burr can pierce the washer and cause a short circuit. Also, a slightly raised edge around a mounting hole can act like a wedge that will split the brittle washer when the component is screwed in place. Thus, every mounting hole and cutout associated with an insulating washer or collar must be completely deburred. Use coarse steel wool to finish the smoothing job. The final test is to rub your finger across the hole. You should feel *only* the depression.

In addition, scrape away the paint on any surface beneath a transistor insulating washer (or other power semiconductor device). This helps heat transfer between the device and the chassis or heat sink. We'll say more about this point later.

Mounting Components

Common sense is your most valuable tool, and you will find the following rules worth noting: Whenever possible, mount heavy components last—the chassis becomes unwieldy very quickly after heavy items are bolted in place. Remember to schedule parts mounting so that components installed early don't block the mounting holes of other parts you plan to install later. Fragile components easily damaged by misguided tools—such as delicate coils—should be near the end of your mounting schedule.

Cut the leads of components like transformers to proper length *before* you mount them—it's hard to work a wire stripper inside a tight chassis.

Follow the hint of kit manufacturers, and plan to prewire (as much as possible) switches and control terminals before you mount the parts. This is especially important if the components are mounted in tight corners where there's little room to swing a soldering iron.

Insulators and/or spacers that support subchassis should raise the lowest protruding point on the subchassis at least ¼ inch above the surface of the main chassis. This will prevent slight chassis metal flexing, subchassis flexing, or chassis dents, from short-circuiting the subchassis.

Always use lockwashers. These diminutive gadgets serve several functions:

• The sharp-edged teeth bite into both the mating surface (the metal area directly below the nut) and the face of the nut, thus holding the nut still while you tighten the screw, and preventing loosening later.

• The washer's springiness introduces a tension force that helps pull the screw head against its mating surface, again preventing loosening later on.

• The teeth dig deep into clean metal, helping to create a good electrical connection for those components that are grounded to the chassis via their mounting lugs.

Normally, a single lockwasher placed directly under the nut is sufficient; here are several situations, though, that call for different washer placement:

• Use two lockwashers to mount each of the lugs of terminal strips. Place one on either side of the lug. The washer between the lug and the panel prevents the relatively flexible lug "arm" from twisting when you tighten the screw; the washer between lug and nut locks the nut in place.

• Fit a single lockwasher under the head of each screw used

to fasten a tapped component. It will prevent screw rotation and loosening.

• Whenever an exceptionally good electrical connection is required, say when you mount a variable capacitor on a chassis, place a lockwasher between the component and chassis at each mounting point.

Do a complete job of mounting panel-mount components. Unfortunately, switches, potentiometers, and similar controls aren't always supplied with all of the required hardware. They usually come with only a shaft nut, and you will have to buy the remaining hardware needed for a proper installation.

The *control-shaft lockwasher* serves the functions described above.

The *control-shaft washer,* placed directly underneath the shaft nut, protects the panel surface beneath the nut from damage (the rough edges of the bare nut will chew into the unprotected surface, leaving a visible ring), and it helps raise the wrench used to tighten the nut well above the surface of the chassis. Tool scratches are a common kind of disfiguration, and are easily prevented by the generous use of washers.

Note: Many panel-mount controls are supplied with a raised "indexing tab" designed to key into a small hole drilled adjacent to the main shaft hole. Its purpose is to speed component orientation during assembly-line manufacture of equipment. Press the tab out of the way, or snap it off with pliers, before you mount the component.

Incidentally, control washers and lockwashers are available from any electronics supply house.

Because panel-mounted fuse holders are mounted with the body nut inside the chassis, they are supplied with an "invisible" lockwasher—a thin black rubber ring—that fits between the holder's front flange and the chassis surface. Be sure you install this ring whenever you mount a fuse holder.

Concentrate on small torque when you handle a screwdriver or wrench. Use just enough force to tighten mounting screws and nuts—excessive torque can damage threaded shaft-mount components, crack plastic mounting holes and lugs used on some components, and strip the threads of machine screws and nuts.

Definitely apply silicone grease (heat-sink compound) to both sides of the insulating washer used between a power semiconductor component and its heat sink (or chassis). The grease fills the tiny surface voids, and helps insure good heat transfer from the device to the sink.

Incidentally, using excessive torque to tighten the mounting nut on the threaded stud of a high-power semiconductor device can lead to overheating during circuit operation. The reason is that the torque deforms the case, and prevents the bottom surface from making an intimate mechanical connection with the heat sink via the washer. A good surface-to-surface contact is required for efficient heat transfer.

Working with perforated board. Cutting a small perforated-board subchassis from a large board (the most economical way to buy this material) can be done two ways:

A sharp "karate chop" will break a perforated phenolic board instantly, and leave a neat edge on both pieces. Be sure that the table underneath has a sharp, not a rounded, edge.

Commercial pusher tool makes it easy to insert push-in terminal in a perforated phenolic circuit board . . .

. . . Or you can use a pair of needlenose pliers. Be sure to apply pressure straight downwards, or you'll bend the terminal.

1. With a fine-toothed saw blade designed to cut plastic. The phenolic material is brittle, so work slowly and evenly, along a line of holes, and use the minimum blade pressure that lets you cut without tooth skipping. Also, use this method to trim a small board to size when you have to remove only one or two "rows" of solid material.

2. By fracturing the board along a line of holes. Place the selected line along—and over—the edge of a sharp-edged table or workbench. Press firmly, with the heel of one hand, on the piece of board resting on the table, and bring your free hand—palm flattened—down smartly against the section of board hanging over the end of the table. Use a wood file to smooth the rough edges of the break.

Several board-shattering experiences have taught me that the best way to drill holes in perforated phenolic is to follow the rules for working with plastic: Enlarge small holes with a reamer, rather than drill large holes directly. Whenever possible, enlarge existing perforations, rather than drill new holes in the solid areas.

Most of the rules for mounting metal chassis components apply equally well when you work with perforated board—with an especially strong warning about using excessive screw tightening torque. One turn too many usually splits the board.

Push-in terminals are components unique to perforated chassis boards, and they introduce a unique problem: How to install the terminals without cracking the board or (more likely) bending the terminal?

Although pusher tools (hollow-end metal rods that hold the soldering end of the terminal) are available, I have found that a pair of longnose pliers work as well when used as shown in the photo. Grip the soldering end of the terminal tightly between the plier jaws, and position the insert end at the top of the intended mounting hole so that it is perpendicular to the board. Push firmly until the terminal locks in place.

Two unusual assembly techniques that you should consider are pop-riveting and the use of epoxy cement.

Rivets can be used instead of screws to install most lug-mounted components. The one disadvantage is that it is much harder to remove a rivet (you must drill it out) than a machine screw. The offsetting advantages, though, are the inherent permanence of a rivet, the relative freedom from unloosening due to vibration or shock, and the excellent electrical connection formed between two riveted-together metal components. Keep these points in mind:

• Don't use rivets to fasten components to perforated board—the riveting force is likely to split the board.

• Similarly don't use rivets to mount plastic components—such as barrier terminal strips.

• Make sure you use the right length rivet for the job. Too long a rivet can't form a good mechanical joint. Eighth-inch rivets are the most oft-needed length.

• Use aluminum rivets to mount components to aluminum chassis. The harder steel rivets chew into the soft aluminum surface when installed, and may not set properly.

Epoxy cement forms a permanent, exceptionally strong bond, that has many potential electronic construction applications. Here are a few examples from my own projects.

• Cementing an aluminum heat sink to a perforated phenolic subchassis.

• Mounting component-supporting brackets inside an aluminum cabinet when it was undesirable, for appearance sake, to have screw heads visible on the cabinet's outer surface.

• Cementing a battery in place in a chassis rather than use a holder. (The circuit power requirements were such that the battery would have a very long service life.) Replacing the battery is fairly simple: Break loose the cement bead (this tears away some of the battery's plastic cover) and cement a new unit in place.

Generally, epoxy cement will bond most common materials used in electronics—including bakelite. Here's a hint: To speed up the setting time, place the cemented components inside an oven set for very low heat (about 170 degrees F). Setting takes a little over an hour with most epoxy formulas. *Note:* Bake the epoxy before any electronic components are installed. Obviously, you can't speed up the cementing of heat-sensitive electronic components, including batteries.

Soldering and Wiring

In any multi-step creative process, there's usually one particular step that most strongly affects the outcome. In building an electronic project, it is wiring and soldering.

The various component leads and connecting wires are the conducting paths that carry current throughout the circuit. And solder binds the leads and wires together. Each solder joint is a secure mechanical *and* electrical bond.

A bad solder joint, a faulty connection, a short-circuited lead, a heat-damaged component, to name a few potential wiring flaws, will stop a circuit dead, no matter how professional a planning and parts-mounting job you've done. And all of this leads directly to the question of soldering.

Actually, soldering is not a particularly difficult skill to develop, so it's surprising that the thought of joining two wires together with a blob of hot solder seems frightening to many people. The basic techniques are straightforward, and important enough to rate lead-off billing in this chapter.

Soldering

Solder is the stuff that binds ·individual electronic components together into functioning electronic circuits. It is roughly analogous to the mortar in a brick wall, or the glue and screws in a piece of furniture, with one important difference: besides forming a strong *mechanical* bond, solder creates and maintains an excellent electrical connection between the components it joins together.

Actually, the name "soldering" is a rather vague term that describes two completely different metal-joining processes. *Hard soldering*, a process that requires very high temperatures, is really a form of brazing. The surfaces of the two pieces of metal that are joined actually melt, and mix together with the molten hard-solder to form a fusion alloy. When the joint cools, the solidified alloy holds the pieces of metal together.

Soft soldering, the process we are concerned with, works on a totally different principle, and soft solder is totally different from hard solder. Solder (we will drop the "soft" from here on) is a very simple alloy of two common metals, tin and lead. It has one amazing talent: when it is molten it can dissolve many other common metals. The other metals don't melt—the temperature of molten solder is far below the melting point of all solderable metals—but quite literally dissolve, in much the same way as table salt dissolves in water. Thus, the joint between two copper wires that have been soldered together consists of a solidified solution of copper in solder.

Soldering is an ideal way to form electrical and electronic connections for several reasons:

• Solder joints can be formed very quickly, and the technique is easy to master. Best of all, the process is inexpensive.

• Solder joints are physically small, and mechanically very strong. They are liquid- and gastight, and relatively corrosion resistant.

• Solder joints maintain excellent electrical connections since they are, essentially, blobs of highly conductive metal.

A wide range of solders is available that differ in their relative proportions of tin and lead. The best solder for electronic wiring contains roughly 60 percent tin and 40 percent lead. To understand why, you must look at the *tin-lead fusion diagram*.

Pure lead melts at a temperature of 621 degrees F; pure tin melts at 450 degrees. Yet when tin is added to lead, or lead is added to tin, the melting points of the new alloys created are much lower. The lowest possible melting point is 361 degrees; this occurs with an alloy of 63 percent tin and 37 percent lead. An interesting point is that this particular alloy, which is called the eutectic composition, is the only one that possesses a distinct melting point. Every other tin-lead mixture goes through a plastic stage during which it is neither fully liquid nor fully solid.

Eutectic composition solder is most desirable for two reasons: Its well-defined melting point eliminates any possible confusion about whether or not the solder is fully melted. This is important since a solder joint can form only when the solder is fully liquid. And comparison tests have shown that eutectic composition solder forms the best solder joints.

Theoretically, a solder joint should form whenever two pieces of solderable metal are heated to about 550 degrees F. Tests have proven this the optimum temperature regardless of the solder's melting point. However, this technique won't work in practice, since all common metals are covered with a thin nonmetallic film of oxides that blocks the molten solder from reaching the metals' surface and forming a metal-solder solution. This film must be removed before a solder joint can be made. That is the function of *soldering flux*.

Flux dissolves the surface oxides, and lifts them away from the surface, so the molten solder can reach the metal. Note that it doesn't play any other part in the solder joint's formation.

Several different fluxes have been developed, but only one class—the rosin or resin fluxes—can be used in electrical wiring. All other fluxes, including the so-called "acid" flux used in large-object soldering, leave harmful residues behind after the joint cools. Besides being corrosive, these residues can absorb moisture from the air and become electrical conductors—they are capable of short-circuiting an electronic circuit.

Rosin or resin fluxes, though, are corrosive only when they are hot and molten. At room temperatures they solidify into almost inert plastic-like masses that are virtually perfect electrical insulators.

To make a simple process even simpler, solder made for electronic circuitry is prepared in a wirelike form with the flux built in—the rosin or resin flux threads through the solder in one or more *cores*. As the solder melts, just the right amount of flux is automatically applied to the joint in liquid form. When the joint solidifies, the rosin hardens on its surface, still carrying the metal oxides it removed.

How to solder. The seven steps that leave you with a perfect solder joint are detailed below. Note that you won't clean and/or retin your soldering iron or gun tip before each joint. You'll usually clean it each time you begin work, and you'll tin it periodically (every 15 minutes or so) as you work. Tinning is explained in the second step, below.

1. Clean the tip before you heat the iron or gun. If the tip is pure copper, use a fine-toothed file to scrape away oxides, cold solder residue, and corrosion. Expose as much clean metal as possible, and remove shallow pits. If the tip is plated, use fine sandpaper to remove surface corrosion and residue, but be careful not to scrape away any surface plating.

2. Tin the tip. Tinning leaves a thin film of molten solder on the tip's surface that helps transfer heat from the iron to the metal surfaces being joined—a poorly tinned tip will not form good solder joints. Tin a pure copper tip as soon as it heats up; if you wait more than a few minutes, the hot copper surface will oxidize and corrode. Tin a plated tip after you wipe off any remaining residue with a damp sponge or a steel-wool pad to expose the clean, plated surface. Here's how:

Apply a liberal amount of solder to all tip surfaces. Run the solder back and forth along the tip so that molten flux will "wet" the whole surface as the solder melts. Shake off excess solder (aim at a damp cloth spread across a corner of your wiring table). Repeat, by adding fresh flux-filled solder, until the entire tip is coated with a silvery solder film.

As you solder, pause every few minutes to brush away residue buildup with steel wool, and examine the tinned surface. Touch up the tinning, if necessary, by adding fresh solder and shaking off the excess as before.

3. Prepare the surfaces to be joined. Remember, flux is designed to cut through oxide coatings—not grease, dirt, dust, wax, or enamel insulation. Unless removed, these substances can prevent molten solder from flowing across the surfaces, and making intimate contact. Fortunately, most of the leads, lugs, and wires you will work with are factory tinned. They are coated with a thin film of solder that protects the base metal from corrosion and makes is easy to solder to them.

Pause every few minutes as you solder and use a steel-wool pad to wipe away residue buildup from the soldering-iron tip.

Most will be bright and clean when you first handle them. If some are not, use an all-purpose solvent to remove grease and dirt; scrape away wax or enamel with a knife edge (be careful not to scrape off or nick the surface tinning).

4. Heat the connection to the correct temperature with the iron. By "connection" we mean the assembly of leads, lugs, wires, etc. that are to be soldered together.

Practically speaking, the "correct temperature" is reached when flux and solder melt almost instantaneously and flow quickly along the surfaces to be joined. The time required for a given connection to reach this temperature depends on the mass of the metal being heated and on the wattage rating and tip size of the iron. It might take one second, or ten seconds, or somewhere in between.

Once you have applied the iron to the connection, re-

Good solder joints are smooth, have no sharp edges, and shine with an overall silvery glow (above). Bad solder joints are grainy, rough-textured, and dull—like fine-grain steel wool (below).

peatedly tap the end of the solder coil against the connection. As soon as the surface "grabs" the solder, and melts its tip, you can assume correct connection temperature.

5. Apply solder sparingly to the connection, by running the end of the solder coil over all of the surfaces to be joined. See that a film of molten flux flows just ahead of the wave of molten solder.

Cover all of the connection surfaces, but don't pile the solder on. The strength of the finished joint will lie in the thin solidified solder film between the various surface pairs. Excess solder doesn't add appreciable strength, but it does cost money, spoil appearance, and may cause short circuits by dripping off the connection.

6. Remove the soldering iron and allow the joint to cool without disturbing the wires and leads. This is very important. Any jiggling or lead motion before the solder solidifies may cause a "cold-soldered" joint. Simply, this means that the chemical "lock" hasn't closed—the joint has poor mechanical strength, and high electrical resistance.

7. Inspect the finished joint carefully. It should have a smooth, shiny, silvery surface, and should bond the leads and wires securely. If the surface is a dull gray color, or looks grainy or rough edged, or if any of the leads are loose, reheat the connection and add more solder.

Also—and this is especially important when you solder to any large mass of metal, such as a steel chassis—check to see that the joint looks like an integral addition to the metal surface. If the solder "blob" looks like a silver drop of water resting on the surface, it probably hasn't made a good joint. (Chances are that a flick of a screwdriver tip will flip it off.) A perfect joint will seem to flow into the metals being joined, with an almost undetectable boundary where the solder ends and the base metal begins. The remedy is to apply heat again, along with a small length of solder. Give the flux from this solder a few seconds to flow and cleanse before you apply additional solder.

Three bad habits. The people who manufacture electronic kits estimate that 90 percent of the assembled kits that don't work, don't work because of faulty solder joints. I would guess that most of these faulty joints are the result of three bad soldering habits:

1. *Not heating the connection uniformly.* This often happens when a small-tip, low-power pencil soldering iron is used to heat a large connection. First, one area is heated and a bit of solder flows; then another area, with some more solder added; and so on until the joint is "finished." Finished is a good word! There's little hope that the various leads and lugs are linked together via a good solder connection.

2. *Heating the solder instead of the joint.* For some reason, many people treat a soldering iron like a paintbrush. Their theory is that it's okay to load the iron's tip with molten solder, and then smear it across the connection like baseboard enamel. Unfortunately, it doesn't work, for two reasons: First, the flux within the solder rapidly disappears in a stream of sweet-smelling smoke—well before it has a chance to clean the connection surfaces. And second (and more important), the molten solder cools almost instantaneously as it is "smeared" on. There is no time for metal surfaces to dissolve.

Remember: A soldering iron is a tool designed to heat the connection to a temperature hot enough to melt solder onto itself.

Use the soldering iron to heat the joint, not the solder. Feed in solder from the other side. It will melt when the joint gets hot.

3. *Forgetting that molten solder is a fluid*, with almost magical behavior. The problem is surface tension, the physical property that makes drops of water and drops of molten solder "bead up" when they strike a nonporous surface. This same surface tension leads to the capillary action that enables molten solder to literally climb uphill, or in any other direction, along a suitable path—the "capillary" formed by two almost-touching metal surfaces that are tinned or have been washed by melted flux.

Green plants use capillary action to suck water up from the ground in thin tubes stacked within their stalks. There are no convenient tubes (the capillary kind) on most chassis. But the fingers of a rotary switch contact, two adjacent leads running to two different lugs on a terminal strip, two close-lying conductors on a printed circuit—all of these are makeshift capillaries that can suck up solder if you carelessly provide a blob of excess solder in the right place. A solder bridge across any of them may ruin components, or cause circuit-killing short circuits.

Solidified flux. Should you remove it? It's not necessary, since only molten flux is corrosive. Once cool, rosin or resin flux transforms into an inert, plastic-like substance, which is an excellent electrical insulator. If you choose to remove it for cosmetic reasons, chip it away with a sharp-pointed awl or pick. As you work, be careful that you don't puncture nearby insulation, or cut connecting leads and wires.

Warning: Never, under any circumstances, use acid flux core solder (or acid flux paste) for electronic wiring. The flux residues are corrosive and hydroscopic. They absorb moisture and create electrically conductive paths that will invariably short-circuit chassis components.

Selecting Hook-Up Wire

Solid conductor wire is fairly rigid, and a length snaked around chassis-mounted components will stay put when both ends are soldered in place. And a bared wire end is easily inserted through a terminal lug. However, solid wire will break if flexed repeatedly, and is substantially weakened if accidentally nicked by a wire stripper.

Solidified flux can be left on a circuit board; you do not have to scrape it away. Once cool, the flux is virtually inert.

Stranded conductor wire is fairly flexible, can carry a proportionately higher current than equal-gauge solid conductor wire, and is less susceptible to nicking and scraping by wire strippers and other tools. But its springy nature makes it difficult to keep long stranded wires neatly "dressed" inside a chassis, and the stripped end of a stranded wire must be tinned to hold the fine strands together, before it can be inserted through a small lug hole easily. Therefore, you should have an assortment of both types on hand. Here is my idea of a good hook-up wire library (buy at least 25-feet of each type and/or gauge listed below):

- 22-gauge solid—for point-to-point wiring of low current (small-signal) stages. Excellent for low-power transistor circuitry.
- 20- or 18-gauge solid—for general point-to-point wiring in any type of chassis. This is probably the wire you'll use most often.
- 20-gauge bare (uninsulated) solid—for short connecting links

(between adjacent terminal strip lugs, for example) in low-power point-to-point wiring, and as low-current "bus bar" on perforated board chassis.

• 16-gauge bare solid—for short connecting links in moderate-power wiring, and as medium-current "bus bar" in general circuit applications.

• 22-gauge stranded—for point-to-point wiring in low-power circuitry where flexibility is required; for example, to connect components mounted on the separate parts of a minibox or utility cabinet.

• 18-gauge stranded—for general chassis wiring where little snaking around components is anticipated, for connecting subchassis and other components that may be moved to the main chassis, and for moderate-power flexible wiring.

• 16-gauge stranded—for higher current chassis wiring, such as filament connections, and leads to transformers and power-supply components.

Hint: Purchase several spools of the general-purpose gauges listed above with different color insulation, and change spools often as you wire. This will simplify any future troubleshooting by making different leads easy to identify.

The Lead Length Problem

Lead length is one of those finicky factors that can unsuspectedly affect circuit performance. Perhaps the classic example is the oft-told tale of the hobbyist who put an experimental circuit together on a breadboard. The circuit worked perfectly. Yet, when he translated the breadboarded circuit into a finished version—complete with neat, squared-off interconnections—it wouldn't work. Why? Because his "neat and square" wiring contained excessive lead length that interfered with circuit operation.

An innocent-looking piece of wire can become a circuit killer in several different ways:

Its inherent inductance (every conductor, no matter how small, has some natural inductance) may effectively alter the circuit design. Remember, every lead and connecting wire is in series with some circuit component.

It may lie close to the chassis, or adjacent to another length of wire, and thus add an unexpected capacitor to the circuit plan—the wire's metal conductor serves as one "plate" of the capacitor. The other wire or chassis acts as the other plate. Of course, the wire's insulation is the dielectric. This stray capacitance may be large enough to upset the circuit. High-frequency circuits are especially sensitive to this kind of accidental harassment.

The wire may act like an antenna, and draw electrical noise or other unwanted interference into the circuit.

As a rule of thumb, therefore, keep all lead and wire lengths as short as possible, consistent with good parts placement. This means using direct point-to-point wire routing, and positioning minor components so that their mounting/connecting leads can be clipped to minimum length. The illustrations show several good/bad examples.

There are a few exceptions to the rule, though:

• Leave about ⅛-inch extra length on the leads of heat-sensitive components (such as transistors and other solid-state parts) to allow room for a clip-on heat sink when you solder them in place.

• Heat-producing minor components must usually be positioned away from other components. A bit of extra lead length gives you shifting ability.

• Allow about 50 percent extra length in wires linking the two or three separable parts of a minibox or cabinet, so that you will be able to lay the parts side by side on your work table without stressing the leads and solder joints.

• For the same reason, allow about 25 percent extra length for the connecting leads to subchassis or major components that may have to be moved if troubleshooting is required.

• Allow "flexing" length for the lead wires connected to flexible components. For example, the center spring-loaded terminal on many panel-mount fuse holders sinks downward into the holder body when the fuse is removed.

• In the same way, protect the leads of fragile components (such as miniature coils) by giving them some flexing length. Then, a carelessly handled tool or misplaced finger won't yank the leads loose.

- Do not route wires or leads over or adjacent to mounting bolts. Future mechanical work—say, removing a major component—could cause tool damage to insulation.

The Mechanical Strength Question

A solder joint, as we have said, provides a mechanical as well as electrical bond. However, there is a substantial dispute afoot as to how much of a connection's mechanical strength the solder blob should be responsible for.

The old-line school of thought insists that a connection should be mechanically sound before solder is applied. This means that every wire is tightly wrapped around or over the solder lug or post it runs to, to make a secure mechanical junction. Then, solder is added to serve as a kind of super sealing wax that keeps the wire(s) in place, and ties them to the lug electrically, with a blob of conductive metal.

The newer school holds that the solder itself should shoulder most of the mechanical burden. Here, wires are simply pushed through their intended terminal lugs, or draped loosely around them, and the solder is applied. The cold solder is thus the mechanical "lock" that secures the leads in place.

Keep in mind that by "older" and "newer" methods we don't mean "bad" and "good." A good mechanical connection was a necessity when components and their leads were large and heavy. Today's miniature components place a negligible strain on a good solder joint, so a substantial mechanical joint isn't required. Proof of this is furnished by almost every printed circuit board. The components rest against the backing side of the board, with the leads protruding through holes in the board, and soldered to the foil side. There's no room or opportunity for a traditional "good mechanical connection."

Whether or not you provide a mechanical joint, then, is a matter of judgment. Probably, heavy leads should be wrapped around terminals; so should the wires leading to a crowded lug (three or more leads). And it's a good idea to secure wires likely to be pulled on—such as the wires linking two cabinet halves together.

Whenever you decide not to wrap a lead, keep this point in mind: If the lead is jiggled as the solder blob cools, the resulting joint will be bad. Thus, if you are unable to prevent component motion, secure the leads before you solder.

To Solder or Not to Solder

As you add components to a chassis, you often route several leads and wires to a single lug. The question is: Should you solder each lead in place as you install it, or should you wait until all of the leads are positioned, and then solder the complete connection?

Kit instruction manuals invariably recommend that you wait before soldering. The reason is that the kit builder (unless he looks ahead) doesn't know, as he performs a given instruction, how many more wires will be routed to a specific lug. Soldering prematurely might clog the lug, preventing later additions, and could confuse the builder into making a mistake.

Sensible? Sure, but the major disadvantage of the approach is that is is difficult to solder a multi-lead connection. Often, the bottommost wires aren't touched by flux and/or solder—a flaw that will sooner or later spoil circuit operation.

That's why I recommend that you solder leads in place as you install them. Treat each lug as a multi-site soldering surface—connect each successive lead to a different side or corner of the lug. This will prevent lug blockage, and will insure that every wire is bonded to the lug with a secure solder joint.

Good Wiring Practices

Each of the following suggestions is based on a "mistake" that I and other project designers have made over the years. Once you've made them "on paper"—by reading and thinking about them—you can be reasonably certain that your projects will work when you flip the power switch.

• The order of wiring should be organized to keep your hands, tools, and especially your soldering iron, out of tight corners and away from vulnerable components, leads, and insulation. As a rule, route, connect, and solder the dangling leads of chassis-mounted major components first. Next, add the various interconnecting wires that will be routed against the inner surface of the chassis or cabinet. Then, mount and solder the minor components. Finally, install and connect any sub-chassis, and add the remaining connecting wires.

• Use a pair of pliers, rather than your fingers, to twist together the stranded leads found attached to most major components with leads. Sweat and body oils from your fingertips can form a barrier on the lead's surface that flux will not penetrate.

• Tin all stranded major component leads before you insert their ends into terminal lugs. Hold the lead tip downwards, place the soldering iron against the twisted conductors, and apply the solder to the end of the lead. Capillary action will draw solder up between the strands and lock them together into a single, easy-to-insert mass.

Solderless Connections

Crimped, solderless connections are a practical way to join spade lugs, screw-mount terminals, and removable wire-end terminals to wires and leads. The connections formed are gastight (and so minimize long-term corrosion problems) and are excellent electrical and mechanical bonds—in some ways superior to solder joints. To prove the point, most of the electrical wiring in your car's electrical system is based on solderless hardware.

The heart of the technique is a crimping tool—a pliers-like device with precisely shaped jaws that are designed to crush the cylindrical terminal structure around the wire or lead, in a very special way. The crimping action actually causes metal to flow, binding the terminal and wire surfaces together in a viselike grip.

There are several potential applications in the projects you build. Use crimp on terminals to "terminate" wires leading to screw-type terminals; use butt-end connectors to join wires coming from two parts of a cabinet or minibox if trouble-shooting demanded they be cut; use them to provide secure ground wire connections that can be bolted to an aluminum chassis where soldering is impossible.

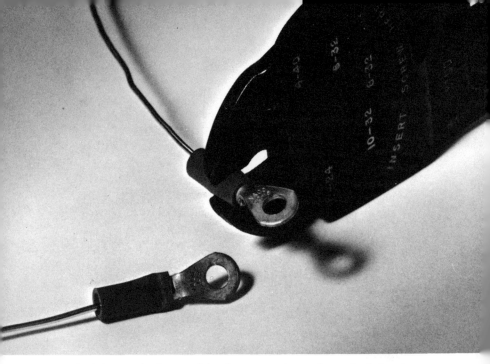

Crimping tools form a joint between wire and terminal that is mechanically secure and electrically sound. Remember to always use the right size terminal to match the wire diameter.

However, be sure you follow the instructions packed with your crimping tool (a bad crimp will fail quickly) and keep these points in mind:

- Use the right-size terminal for the wire you are joining it to—a too-large terminal won't crimp properly around the wire.
- Apply only enough crimping pressure to form a good crimp—excessive force is likely to cut through the terminal, severely weakening the bond.
- Never tin the end of a stranded cable before crimping on a terminal—the solder you add has sufficiently different physical properties to upset the carefully engineered crimp. You will find that the terminals hold perfectly well on stranded wire without tinning.

Getting a Professional Look

Almost as important as building an electronic project that works, is building a project that has a professional look to it. By "professional" we don't mean store-bought or commercially built. There are many reasons why the projects you build at home can never look like commercially built gear. For example, no amateur has access to the elaborate metal-forming machines that produce the custom-made chassis and cabinets used in factory-built equipment. Nor does the hobbyist have the engraving, silk-screen labeling, or fancy spray-painting equipment needed to decorate commercial electronic equipment.

But the projects you build can have the polish and appearance of prototype electronic equipment built by professional electronic technicians. By and large, you work with the same kinds of chassis, cabinets, and components as do the pros. Usually, the difference in appearance between a home-built project and a technician-built device is caused by the hobbyist's lack of attention to detail. It's the small items in decor, trim, and assembly that add to or subtract from good looks.

In this chapter we will discuss the most important elements of a professional look. The majority are low-cost factors that you can incorporate in every project you build; a few add significantly to the cost of the finished project—use these for your showpiece projects.

Custom Paint Jobs

The "hammertone gray" finish of prepainted aluminum cabinets and cases is a serviceable but dull color. And the black crackle-tone finish on prepainted steel cases is a rather old-fashioned decor. Happily, the availability of easy-to-use aerosol-canned spray paints makes it easy to custom finish your projects.

It is difficult to create a really good-looking finish by painting over an existing hammertone or crackle paint job, so start from scratch with a bare metal cabinet. Complete all of the metalworking operations before you begin to paint. The best—and longest-lasting—finish on a metal surface is made by building it of several thin coats, rather than a single thick coat. The reason is that the metal surface is nonporous; the paint layer must dry from the top inwards. The surface film that forms quickly atop a thick layer of drying paint prevents the paint below from drying properly and forming a good bond with the metal surface.

For best results, the first coat should be a metal primer sprayed on very thinly. The correct technique, incidentally, when using aerosol spray paints, is to keep the spray moving quickly over the surface. Lingering on one spot for even a moment will build up a thick, runny paint layer.

I have found that two succeeding color layers make a fine finish on steel cabinets; aluminum surfaces require three (or sometimes four) coats. The reason is that aluminum is a more flexible metal, and a thicker paint film is required to resist chipping and cracking due to surface flexing.

The choice of color is, of course, a matter of taste and intended application of the finished device. As a rule, though, the darker shades of pastel green, tan, blue, and gray make the best colors for electronic gear. These are the colors most often used on commercially built and kit-built equipment.

Two-tone color schemes are becoming increasingly popular. Here, the two halves of a minibox, or the panel and body of a utility case, are each painted a different shade of the same color—light and dark green, for example.

Another approach (and a very attractive one) is to paint a segment of a front panel or cabinet front in a contrasting color. To do this, use painter's masking tape to shield the panel areas that will remain the original color, and apply two or three very thin second-color coats to the exposed surfaces. Allow the paint to dry completely before you remove the tape, to prevent chipping the edges of the contrasting color region.

Satin Finish for Aluminum

This is the kind of finish found on the exposed metal surfaces of a good-quality camera, and it makes an ideal and exceptionally attractive finish for small aluminum-cased projects (especially for portable or hand-held electronic devices). It can be used as well for the aluminum front-panels of projects wired into bakelite instrument cases.

Satin finishing is a chemical process. The aluminum surface is etched by a caustic solution to produce a silky-smooth, slightly granular-appearing finish. It is important that you complete all metalworking operations before treating the aluminum component, since tool marks and scratches become annoying blemishes on a satin-finished surface. The finishing process is very simple. However, the hazards of working with the caustic solution demand that you exercise extreme caution at all times. Wear safety glasses to prevent splashes from getting into your eyes.

Use a polyethelene washtub or enameled basin as the processing tank. Make the caustic solution by dissolving household lye (sodium hydroxide) in *cold* water—about one-third can of lye to a gallon of water. Be sure to wear rubber gloves when you work with the lye or the solution. Use a wooden or hard-rubber stick or implement to stir the solution, and be careful not to splash any solution on your clothing or skin.

The aluminum surface to be finished must be completely clean before it is immersed in the solution. Dirt, grease, or even body oil applied in the form of fingerprints will cause uneven surface etching. Thus, it is a good idea to clean the surface with a solvent such as acetone or alcohol, and rinse it off *completely* with cold running water.

To begin the process, simply immerse the aluminum part completely in the solution. *Note:* Almost as soon as the aluminum contacts the solution, intense bubbling will begin on the exposed surface. This bubbling is caused by a gas given off as a result of the chemical reaction. Be sure that you work in an adequately ventilated area to permit the gas to escape safely.

The correct processing time depends on several factors including solution strength and the degree of surface etching you desire. Typically, it will be between 30 and 40 minutes. However, I suggest that you make a few trial runs with scrap aluminum to perfect your etching technique. (Always renew the solution after each etching "run.")

When removed from the solution (use plastic tongs), the aluminum part will be coated with a fine-particled black deposit. Rinse it off in cold running water, and use a rag to wipe away the deposit.

Note: Because the chemical etching process actually removes aluminum, you will find that the various drilled holes and cut openings are slightly larger after the part is etched. Normally, this slight increase doesn't affect parts mounting. However, if hole dimensions are critical, make the holes slightly smaller (about $\frac{1}{32}$ inch) when you drill and/or cut them. After the part is etched, you may have to touch up hole diameter and opening dimensions slightly with a tapered reamer or a file. Be sure to work neatly and carefully.

Striping

Thin stripes of contrasting color applied to the front panel of an electronic project serve two functions: First, they decorate the panel. This is an especially useful way to decorate a large piece of equipment that has only one or two front panel controls. Second, the stripes can be used to delineate the different groups of controls on the front panel of a complex project. The controls for different functions are grouped within "boxes" made of thin stripes.

Although commercial paint-striping machines are sold in art-supply shops, I have found a much simpler way to make stripes: Use the very narrow plastic tape sold in stationery and art-supply stores. To do a neat striping job with tape, you'll need a pair of sharp-point tweezers, and a small-blade hobby knife (X-acto or equivalent).

How to Make Labels

There's little doubt that ugly labels have ruined the appearance of more home-built projects than any other single factor. To begin with, hobbyists are often tempted to overlabel—that is, they use more labels than are necessary to explain the function of each panel control. And then, too many hobbyists use messy labeling techniques, such as writing on adhesive tape, or (the worst possible way) scratching the labels into the paintwork with an awl or an ice-pick.

Commercial equipment is usually labeled either by engraving the legends on the front panel, or applying them via a silk-screen printing technique. Neither is suitable for home-workshop application. However, you can effectively simulate the appearance of silk-screened labels with transfer decals (available from most electronic supply houses) and with press-on transfer type (sold in most art-supply stores).

Decals are available in a wide variety of control legends and assorted miscellaneous panel markings (arrows, curved lines, 1-to-10 scales for use under control knobs, etc.). Red, dark gray, white, and black are the standard colors. Usually, you must purchase a set of labels that includes several hundred different individual decals. Normally, popular legends, such as "power," or "on" and "off," are repeated several times on each sheet.

Decals are surprisingly durable, if applied according to directions and coated, when dry, with a clear plastic spray. Besides toughening the decal's surface, the spray helps keep the clear connecting film (the film that ties the letters together) from darkening with age, and ruining the label's appearance.

The standard legends supplied on a sheet of decals can be used to create unusual words or phrases by cutting the stock words apart and recombining the pieces. Be sure you maintain the original inter-letter spacing, and position the pieces carefully so that their base line isn't ragged.

Transfer type is widely used by commerical artists, and one of its great appeals is that it is available in a very wide range of type faces and type sizes and colors. It consists of a set of waxlike letters that are temporarily mounted on waxpaper. You create any word you want by transferring one letter at a time to the panel's surface. This is done simply by laying the rear surface of the letter against the panel, and then burnishing the waxpaper directly above the letter. The letter leaves

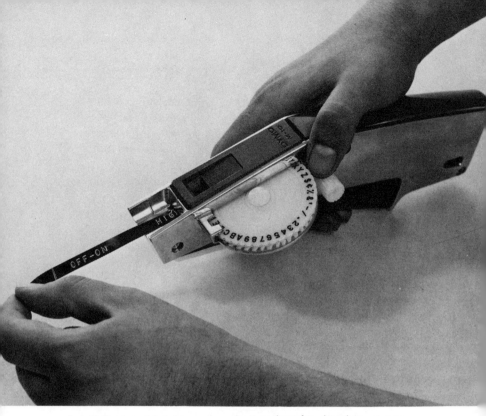

A labeling machine is a convenient gadget for dressing up any project. Select a label color that contrasts with the cabinet color.

the sheet, and transfers to the panel's surface.

One disadvantage of transfer type is that the wax letters are relatively fragile; a careless swipe of a fingernail can scratch them, so plan to use the material on panel areas that receive the least finger traffic.

Label-making machines (such as the DYMO) produce highly legible—if somewhat tacked-on looking—labels. Actually, you can tone down the impression that they were added as an afterthought by carefully choosing the color of the labeling tape used, and by incorporating the labels into wide color strips. The first point simply means that the color of the tape should not clash with the color of the panel; ideally it should complement it. And, second, plan the location of the labels so that blank areas of tape form color bands or strips on the panel. One final point: This type of label should not be used on a cluttered panel; it looks good only when applied to a sparsely instrumented panel.

Customizing Panel Meter Scales

Almost all of the panel meters you will work with will have basic scale calibrations: 0-to-1 milliampere; 0-to-5 volts DC; 0-to-200 volts AC, for example. The two most important exceptions are the VU-meter, and the S-meter—both are equipped with special-purpose scales.

However, most of the projects you will build that use meters will require that the meters have nonstandard meter scales. For example, an automotive tachometer may require a meter that reads 0-to-5,000 R.P.M., even though the meter movement is really a familiar 0-to-1 ma milliammeter.

Altering a meter scale is a delicate—but straightforward—job. The most successful technique is to leave the basic scale markings alone, and concentrate your efforts on relabeling the

Homemade panel meter scales are a necessity for giving any project a finished look. Be careful: the meter's moving-coil mechanism is fragile, so don't pull or bend the needle as you mount the new scale. Also, when you reinstall the front of the housing, make sure that the zero adjusting cam in the front piece lines up with the adjusting tongue on the movement.

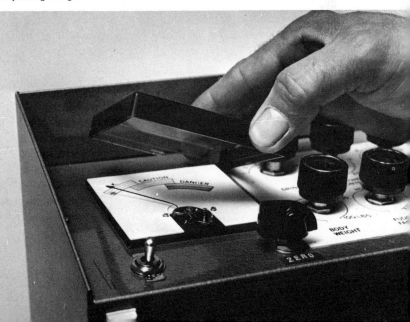

various numerical calibrations to fit the new application. Although some experimenters create an entirely new meter scale—including the scale markings—this is a very difficult job that requires considerable drafting experience.

The first step is to remove the meter's protective glass or plastic face. This simply snaps off, in many modern plastic-cased units, but older glass-faced meters often have front pieces that are bolted to the main body.

Next, cut a crescent of thin but opaque drawing paper just large enough to cover the existing numerical calibrations. The thinness of the paper is important, since it must not interfere with the movement of the meter needle. Mount the crescent in place with a few dabs of a plastic-based adhesive (such as Elmer's Glueall). Keep the glue film as thin as possible.

Cut and mount, in the same way, another paper crescent to cover the meter's units legend (DC Volts, or DC Milliamperes, or AC Amps, etc.)

Be very careful not to disturb or deform the meter needle as you mount the paper crescents. You'll find that, with the help of a pair of tweezers, you can slide the paper slips in place underneath the needle without moving it. Also, be sure you don't drip glue into the movement.

Allow the glue to set thoroughly, and then, using a fine-tipped lettering pen and india ink, write the new numerical calibrations in the appropriate places on the calibration crescent. Or, instead of pen and ink, use transfer-type numerals and burnish them in place, being careful not to damage the needle with your burnishing tool.

Finally, reinstall the faceplate. Before you press the faceplate into place, check to see that the zero-adjusting cam mounted through the faceplate mates properly with the zero-adjuster tongue on the meter movement.

Accessory Hardware

There are a variety of hardware items which, though nonessential to circuit operation, serve to complete a project.

Rubber feet. Most types bolt to the base or bottom surface of the cabinet; newer types mount via a contact adhesive.

Metal handles. Plated metal handles can be mounted on top of the chassis to serve as carrying handles, or on opposite sides of the front panel to serve as "roll bars," an excellent way to protect fragile meter faces and plastic knobs on portable test instruments.

Line-cord strain relief. This simple plastic gadget is more effective than a knot tied in the cord, and a good deal neater.

Decorative external screws and bolts. Take a tip from commercial gear and use black or gold bolts wherever these fasteners will be visible—for example, to mount parts on the inside surface of the front panel.

Switch plates (dial plates). These fit beneath the knobs of rotary switches and potentiometers to provide indexing and positioning scales. They are often neater than hand-drawn or decal scales.

High-quality knobs and fittings. The simple and cheap plastic knobs often supplied with rotary switches and potentiometers are fine, most of the time. But for really topnotch appearance, use machined metal or high-quality molded plastic knobs. They are expensive, but if good looks are important they are worth the cost.

In anything electronic, beauty is far more than skin deep. The essence of a professional project is neat, tidy, circuit wiring. Here are a few professional touches you can incorporate to improve circuit appearance:

• Use cable clamps to secure loose wires and cables to the chassis.

• Use flexible cable grips (ties) and/or lacing cord to tie groups of adjacent wires together into harnessed bundles.

• Keep all solder joints as small as possible—superfluous solder doesn't add to the strength of a joint.

What You Should Know About Test Instruments

Talk about electronics, and most people conjure up a picture of a shelf full of meters, oscilloscopes, and other electronic test paraphernalia. It's only natural, since the right test gear is as important to the man who works with electronics as a square and spirit level are to a cabinetmaker.

Unhappily, the word "test" suggests that test instruments spend most of their time troubleshooting out-of-order electronic circuitry. This just isn't true. You will call on your test gear to help you perform a wide variety of electronic odd-jobs; troubleshooting is only one item on the list. Here are some others:

Measurement. What is the resistance of an unmarked carbon resistor? What is the output voltage of an elderly battery? Is an electrolytic capacitor "open," or is it "shorted"? What is the frequency response of a stereo amplifier? These are questions that can be answered only by taking measurements with the appropriate test equipment.

Circuit adjustment. Some people call it "fiddling with the controls," but the basic idea is to adjust various adjustable components within a circuit so that the circuit works properly. For example, many amplifier circuits include variable resistors in their output stages that must be adjusted to establish the correct "bias" currents or voltages. The hard way to do this job is by trial and error; the easy way is with the aid of a DC ammeter (current measuring meter) or DC voltmeter temporarily inserted into the circuit.

Monitoring performance. Think of the different gauges and signal lights on your car's instrument panel. These are really simple test instruments that watch over the engine's performance. In much the same way, electronic equipment often has built-in test instruments to monitor circuit operation, or has provision for the temporary use of external test equipment.

A good example of a built-in instrument is the VU meter (volume level meter) found in many tape recorders. This is basically an AC voltmeter that tells the operator what the machine's recording amplifier is doing.

Externally speaking, you will make many different checks from time to time. One example: Checking the output voltage of a power supply with a DC voltmeter to make sure it's up to specification.

Troubleshooting. This is as much of an art as it is a science. Finding the faulty component, or bad solder joint, or short circuit, or wiring error that has laid low an electronic device usually requires the kind of sleuthing talents made famous by Sherlock Holmes. But test instruments can give you a helpful head start as you search:

Continuity tests, made with an ohmmeter (device that measures electrical resistance) or a continuity checker (we'll discuss this instrument shortly), will point out short and/or open circuits in chassis wiring.

Resistance tests, again made with an ohmmeter, will tell you if the correct resistance values exist between each component terminal and the circuit ground (the common component lead-connection point, often the metal chassis).

Voltage tests, made with a voltmeter, can verify whether or not the correct voltage levels exist between the various component terminals and circuit ground.

Test Instruments You Should Own

There are almost as many test instruments available as there are things to measure and check and test in an electronic circuit. Fortunately, as with wiring tools, you don't have to own more than a basic few.

The four test instruments that I consider essential are the *continuity checker,* the *neon-bulb voltage indicator,* the multi-range *volt-ohm-milliammeter* (or VOM), and the multi-range *vacuum tube voltmeter* (or VTVM).

The first two instruments are extremely simple gadgets that cost about a dollar each. Both the VOM and VTVM are available either factory wired or in ready-to-wire kit form. Kits cost about $30 each; the factory-wired versions cost about 50 percent more.

I suggest that you consider purchasing at least one of these important test devices in kit form, even if you have no prior kit-building experience. Thousands of fledgling electronic buffs have cut their kit-wiring teeth by assembling their own VOM or VTVM.

We'll discuss these four basic instruments, and their operating fundamentals, in this chapter. Later in this book, we will talk about several of the other test instruments you will see on the pages of an electronic supply-house catalog. The chances are good that you will eventually acquire one or more of these specialized electronic tools as your interest and experience in electronics grow.

Continuity Checker

Occasionally called a circuit tester, this device consists of a low-voltage incandescent bulb, one or two flashlight cells, and a pair of test-prod or alligator-clip-equipped test leads, all wired in series.

Think of your continuity checker as an incomplete flashlight—everything is there but the on-off switch. So, if you touch the test leads together, either directly, or through a low-resistance path, the checker's simple series circuit is completed, and the bulb lights.

You can use the device to verify that wiring is continuous; to check that a switch is working properly; to make sure that the insulated portions of chassis-mounted components are in fact insulated from the chassis; to identify the various wires in a multi-conductor cable; to check whether or not a fuse has blown—the list of uses is almost endless.

To check continuity, first touch (or connect) one of the test

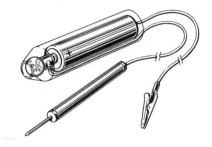

CONTINUITY CHECKER

leads to one side, end, or terminal of the wire, component, or circuit whose continuity is in doubt. Then touch the free test lead to the other side, end, or terminal, and watch the light bulb.

Keep these points in mind when you use the device:
- The checker circuit can not distinguish between an open circuit and a current path whose electrical resistance is greater than 10 or 20 ohms. This slight resistance, in series with the bulb, will prevent the bulb from lighting.
- The relatively high direct current—over $1/10$ ampere—flowing through the checker when the bulb is lit can fatally damage many semiconductor components. Never test the

CONTINUITY CHECKER
SHOWS BLOWN FUSE

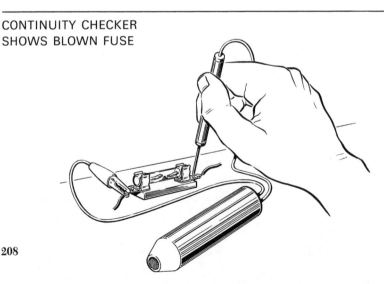

208

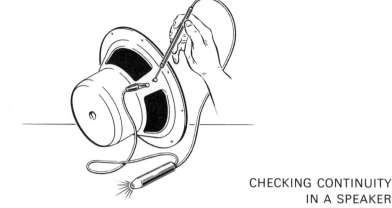

continuity of a diode or transistor, and to be safe, don't check
the immediate wiring surrounding these components. This
same warning holds when you work with other delicate components, such as miniature coils wound of very small-gauge
wire, so plan to use your checker primarily to test high-
current carrying items: chassis wiring, switches, relay contacts, fuses, power transformer windings.

• *Never* check the circuitry of a device when it is operating,
or if its power cord is plugged into an AC outlet. To begin
with, you risk a painful shock, since neither the checker nor
its test leads are insulated for high-voltage testing. Also, the
bulb and battery(s) will be ruined if you inadvertently connect
the test leads across a high voltage. And there's a good chance
that the checker's low resistance (at least before the bulb
burns out) will act as a short circuit, and consequently damage many of the device's components.

Neon-Bulb Voltage Indicator

The neon-bulb voltage indicator (occasionally called a neon
glow tester) is a miniature neon bulb wired in series with a
current-limiting resistor and a pair of high-voltage insulated
test leads. The one you buy will probably look like a plastic
fountain pen, with the neon bulb visible at one end, and two
stubby test leads sticking out of the other.

In its store-bought form, the gadget makes a top-notch
electrical-appliance tester (its original purpose), but its test

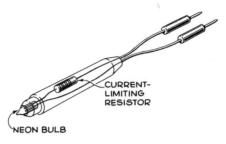

NEON-BULB VOLTAGE INDICATOR

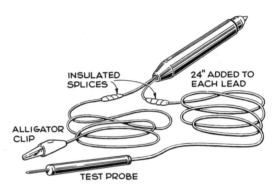

MODIFIED TESTER

leads are too short for safe use under a live electronic chassis.

To lengthen the leads, first clip off the two metal prods. Then, splice on two 24-inch lengths of high-voltage-rated (5000 volts) rubber-insulated test lead cable. Slide short pieces of insulating tubing over the splices, and wrap them tightly with plastic electrical tape. Finish by soldering an insulated alligator clip on one lead, and a high-voltage-rated test probe on the other.

As its name suggests, this device indicates the presence of a high voltage. It is not a substitute for a voltmeter; rather, it provides a safe, convenient, and fast way of determining if high voltage—between 65 and 600 volts AC *or* DC—is present at selected points throughout an operating chassis.

Its heart, the neon bulb, consists of a tiny glass envelope filled with neon gas, at moderately low pressure, and

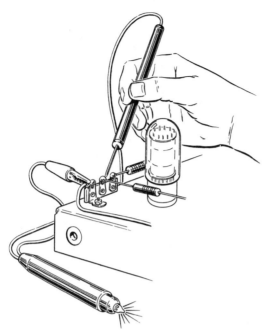

INDICATING PRESENCE OF HIGH VOLTAGE

equipped with two closely spaced metal electrodes. Normally, the bulb has a very high internal resistance: several million ohms. However, if a voltage—either DC or peak AC—of greater than approximately 60 volts is applied across its leads, the neon gas between the two electrodes ionizes, and instantaneously becomes a relatively low-resistance path (about 5000 ohms), allowing current to flow through the bulb. As current flows, the exited neon atoms emit their characteristic red-orange glow.

A lone neon bulb, equipped with test leads, would make a poor voltage indicator because the bulb would carry excessive current and destroy itself, if connected across a high-voltage source. That's why the neon tester includes a current-limiting resistor that keeps the bulb current down to a safe level.

Interestingly enough, a quick glance at the glowing bulb will tell you if the high voltage you are probing is AC or DC. This is because the neon glow surrounds only one of the two electrodes at any given instant: the electrode that leads to the negative terminal of the voltage source.

When the tester is connected to a DC voltage, one electrode glows continuously. When it is connected to an AC voltage, though, the glow shifts back and forth, 60 times each second, between the two electrodes, in perfect step with the alternating electrode polarity. Because your eye can't detect motion at this rate, both electrodes seem to glow steadily.

Keep this simple rule in mind: one electrode glowing means DC; two electrodes glowing means AC.

In most tests, you will be searching for the presence of high voltage between some specific terminal point under a chassis and circuit ground, or between two selected terminal points. The safe test procedure is first to attach the alligator clip to one of the points, or to a suitable circuit ground point (if you are measuring a voltage between a point and ground), and then touch the probe to the other point.

Remember: an operating chassis always represents a dangerous shock hazard when you service or test it. Be sure to follow the commonsense safety rules outlined in the Introduction when you use your neon tester.

And remember that the device is rated for a maximum applied voltage of 600 volts. Never use it to check circuitry where voltages higher than this are likely to be found.

The Basic Moving-Coil Meter Movement

Almost everywhere you look in electronics, you see a panel meter that indicates something. Your vacuum-tube voltmeter will have one; so will your volt-ohm-milliammeter. There's probably at least one meter in your FM tuner to help you tune it correctly. And the electronic tachometer in your car is built around a meter.

Although, the variety of different meters is staggering, virtually all of the meters used in electronic devices have the same kind of heart beating within them: the moving-coil movement.

The most widely used moving-coil movement—often called the D'Arsonval movement after its inventor—consists of a coil

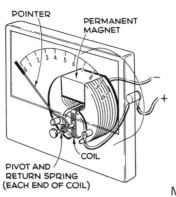

POINTER

PERMANENT MAGNET

COIL

PIVOT AND
RETURN SPRING
(EACH END OF COIL)

TYPICAL METER USING
MOVING-COIL MOVEMENT

of fine wire suspended between two low-friction pivots, and surrounded by a permanent magnet. The movement's pointer—or meter needle—is attached to the coil. When a direct current flows through the coil, it transforms the tiny spool of wire into a miniature electromagnet. The weak magnetic field that the energized coil produces interacts with the magnetic field of the permanent magnet, and generates a small torque—or twisting force—that turns the coil and swings the pointer.

Two tiny hairsprings—one on each end of the coil—oppose the torque and try to turn the coil back to its original position. As a result, the coil—and, consequently, the pointer—move only a short distance, depending on how large a current is flowing through the coil. Most meters are designed so that this current-movement relationship is linear. The greater the current through the coil, the more the pointer deflects.

The rated sensitivity of a moving-coil meter—the amount of current through the coil needed to deflect the pointer to its maximum position—depends on several factors: the number of turns of wire in the coil; the stiffness of the two hairsprings; the weight of the moving parts; and the quality of the pivot surfaces.

With great care and expense it's possible to build a moving-coil meter with a full-scale sensitivity of one microampere DC. This means that a current of only one-millionth of an ampere will deflect its pointer fully.

Note that a moving-coil meter is inherently a direct-current measuring instrument; it won't work on AC. An alternating current fed through the coil would create an alternating magnetic field around it, and the twisting force acting on the coil would change direction in step with the alternating positive and negative current peaks. Consequently, the pointer would either vibrate slightly, or (more likely) seem to stand perfectly still.

The Volt-Ohm-Milliammeter

The volt-ohm-milliammeter is a multi-purpose test meter that measures AC and DC voltages, direct current, and electrical resistance, in several different ranges. That's quite a mouthful, but the VOM is quite an instrument. Without a doubt, it is the measuring instrument you will use most often.

The heart of nearly all popular VOMs is a moving-coil movement with a full-scale sensitivity of 50 microamperes DC. A few expensive laboratory instruments have more sensitive meters, while a handful of low-cost utility VOMs use less sensitive movements—usually one-milliampere (one thousandth of an ampere) DC full-scale meters. The latter, although fine for automotive electrical-system servicing and appliance repair, are not really suitable for electronic circuit testing. We'll see why later.

The volt-ohm-milliammeter's face is covered with several different voltage, current, and resistance scales, arranged in a series of concentric arcs, and usually printed in different colors, so that it is easy to tell which scale is which. Of course, a few markings on its face don't change the meter movement's basic nature, and this brings up the question: How can a meter that indicates current flow also indicate voltage and measure resistance?

The answer is that a VOM also contains a set of clever, but surprisingly simple, circuits that convert AC and DC voltages, and electrical resistance values, into tiny direct currents that the microammeter can measure.

A set of switches—usually including a master, many-posi-

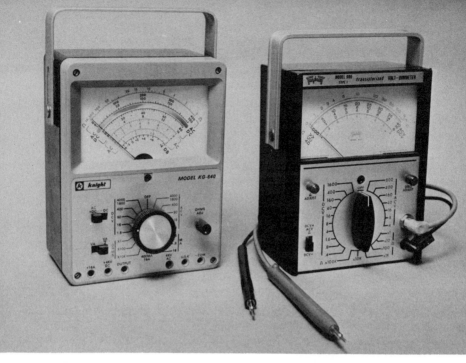

Two vital test instruments are the popular volt-ohm-milliammeter (VOM), left, and the transistorized voltmeter (TVM). The TVM is the modern equivalent of the long-used vacuum-tube voltmeter (VTVM) described in the text.

tion, rotary switch—ties the different circuits to the meter movement. When you select a specific voltage, current, or resistance-measuring range by turning the switch, you actually connect a specific circuit between the meter movement and the VOM's pair of test leads.

Ohm's Law

The key to understanding the different measuring circuits in a VOM is to first understand the most basic relationship of electrical circuit theory: Ohm's Law. So let's go back in time, briefly, to 1827, and look at the researches of Georg Ohm, an obscure German mathematician.

Ohm performed a painstaking series of experiments in which he passed electric currents through various conductors. He observed that when wires made of different materials were connected across the terminals of a Voltaic Pile—a primitive "dry-cell" battery—different quantities of electric current moved through them. He observed the same effect when he tried different size and length wires made of the same material.

From these tests, Ohm deduced that every electrical conductor has a characteristic resistance—or opposition—to the flow of electricity; and that the resistance of a particular conductor depends on its shape, size, and the substance from which it is made. For example, if a long, thin piece of iron wire is connected to a battery, a relatively small current will flow through it; but if a short piece of thick copper wire is connected across the terminals, a very substantial current will flow.

Ohm condensed his observations into a simple relationship:

$$\frac{\text{Resistance of}}{\text{conductor}} = \frac{\text{Voltage applied across conductor}}{\text{Current flowing through conductor}}$$

FIRST FORM OF OHM'S LAW: $R = \dfrac{E}{I}$

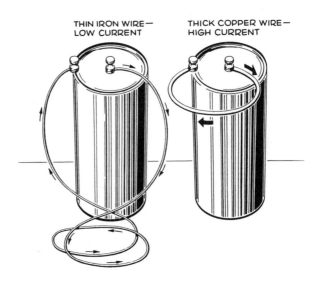

THIN IRON WIRE—
LOW CURRENT

THICK COPPER WIRE—
HIGH CURRENT

In terms of the electrical units we use commonly, Ohm's Law becomes:

$$\text{OHMS (resistance)} = \frac{\text{VOLTS (voltage)}}{\text{AMPERES (current)}}$$

And, finally, in terms of the familiar symbols for resistance, voltage, and current, Ohm's Law reads:

$$R \text{ (ohms)} = \frac{E \text{ (volts)}}{I \text{ (amperes)}}$$

Because this formula is nothing but a simple algebraic relationship, its three symbols can be juggled around to make two other—but totally equivalent—formulas. When we talk of Ohm's Law, therefore, we can mean the formula given above, or either of the alternate forms given below:

$$I \text{ (amperes)} = \frac{E \text{ (volts)}}{R \text{ (ohms)}}$$

This relationship states that if a voltage of E volts is connected across a conductor whose resistance is R ohms, then a current of I amperes will flow through the conductor. And:

$$E \text{ (volts)} = I \text{ (amperes)} \times R \text{ (ohms)}$$

This form of the law states that if a current of I amperes is flowing through a conductor whose resistance is R ohms, then there is a voltage of E volts connected across the conductor.

A Circuit to Measure DC Voltage

If you place a resistor across a source of DC voltage, a current will flow through the resistor. The value of the current is given by the second form of Ohm's Law: I =

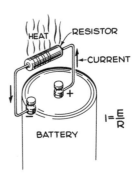

SECOND FORM OF OHM'S LAW: $I = \dfrac{E}{R}$

$E \div R$. This is a linear relationship, so, if you double the voltage across the resistor, the current flowing through it will double; if you halve the voltage, the current will be cut in half.

Add a microammeter in series with the resistor, and you have a rudimentary voltage-measuring instrument—a simple voltmeter. This is because for any particular voltage you apply across the resistor/microammeter pair, a particular current—as specified by Ohm's Law—must flow through the pair, and be registered on the meter's face.

Looking at it the other way: A particular current reading on the meter means that there is a particular voltage applied across the resistor/microammeter pair.

By selecting an appropriate value resistor, and an appropriate meter sensitivity, we can tailor the finished voltmeter to read any desired voltage range.

As an example, let's design a voltmeter that will measure any DC voltage between 0 and 10 volts. We will use the same type of meter movement found in a VOM: a 50-microampere sensitivity microammeter.

The first step is to divide the meter's face into ten equal segments, and print a scale labeled 0—1—2—3—4—5—6—7—8—9—10 on the face just below the pointer's tip. Thus, the pointer will rise to 10 when a current of 50 microamperes flows through the meter, and it will stop at a proportionately lower reading with a proportionately smaller current. For example, a current of 20 microamperes will cause a reading of 4; 45 microamperes produces a reading of 9.

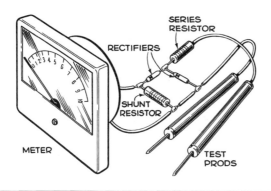

Next, we must select the correct value series resistor. Since we want a voltage of 10 volts to deflect the meter needle fully (to the 10 on the scale), we must choose a resistor value that will allow 50 microamperes to flow through the resistor/microammeter circuit when 10 volts DC is applied across the pair. Turning to the first form of Ohm's Law:

$$R = \frac{E}{I}$$

$$R = \frac{10}{.000050} = 200,000 \text{ ohms}$$

Thus, the *total* resistance of the resistor plus the microammeter wired in series should be 200,000 ohms.

Normally, the internal resistance of a 50-microampere meter movement is a few hundred ohms or less. This is, in effect, the resistance of its coil of fine-gauge wire. Less sensitive meters require fewer turns on their coils, and hence have lower internal resistance values. We will assume that our meter's resistance is 300 ohms.

Since the total resistance of two resistors wired in series (the meter's coil acts like an ordinary resistor in the circuit) is just the sum of the individual resistances:

$$R_{total} = 300 \text{ ohms} + R_{resistor}$$

$$200,000 \text{ ohms} = 300 \text{ ohms} + R_{resistor}$$

therefore

$$R_{resistor} = 200,000 - 300 = 199,700 \text{ ohms}$$

Because the difference between 200,000 ohms and 199,700 ohms is so slight—much less than the 1 percent variation implied in the resistor's tolerance value—we can safely neglect the meter's internal resistance, and use a 200,000-ohm resistor in the circuit.

The following table proves that this simple voltage-measuring circuit really works:

DC *Voltage* applied to test leads (*volts*)	Approximate direct current through meter (*microamperes*)	Reading on meter's 0-to-10 scale
2	10	2
6.5	32.5	6.5
4.3	21.5	4.3
10	50	10
3.9	19.5	3.9

The DC voltage-measuring circuits in all VOMs are based on the simple series resistor/microammeter circuit described above. Of course, if you look inside a VOM, you will find many more components at work in the voltmeter stages. The reason is that a VOM's voltage-reading circuits are designed for multi-range operation. Popular VOMs have anywhere from four to ten (and sometimes more) overlapping DC voltage ranges. A typical unit, for example, might have seven ranges: 0-to-1 volt; 0-to-5 volts; 0-to-10 volts; 0-to-50 volts; 0-to-100 volts; 0-to-500 volts; and 0-to-1000 volts.

Why so many different ranges? The reason is so that you will have an appropriate voltmeter at your disposal to measure any of the widely ranging voltages you can encounter in electronic equipment. It's obvious that you can't use a low-voltage voltmeter to measure a high DC voltage—the meter movement and/or the series resistor will be damaged if you try. But, on the other hand, you shouldn't use a high-voltage voltmeter to measure a low DC voltage—it's not accurate enough.

If, for example, you measure the voltage of a 1½-volt dry cell with a 0-to-100-volt DC voltmeter, the meter's pointer would certainly move. But the movement would be so slight as compared to the meter's scale markings, that it would be extremely difficult to read the indicated voltage accurately. More important, though, is the problem of the meter circuit's inherent accuracy. This is usually specified in terms of a percentage value of the meter's full scale reading. Thus, if a 0-to-100-volt DC voltmeter is guaranteed accurate to within 3 percent, this means that any voltage measurement made with it is accurate to within 3 percent of 100 volts, or 3 volts. (*Note:* The DC voltage ranges of virtually all moderately priced VOMs have a rated accuracy of plus or minus 3 percent of full scale value. Therefore, a reading taken on a 0-to-1-volt range will be accurate to within .03 volts; a reading taken on a 0-to-1000-volt range will be accurate to within plus or minus 30 volts.)

You've probably guessed by now why it doesn't make much sense to measure a 1½-volt battery with a 0-to-100-volt DC voltmeter. The meter's allowable inaccuracy—plus or minus 3 volts—is greater than the voltage you are trying to measure. Consequently, you can't possibly believe the reading you see.

The heart of a multi-range voltmeter circuit is a multi-valued series resistor. Changing the resistor's resistance value changes the meter's full-scale voltage calibration. In all popular VOMs, this multi-valued resistor is actually a chain of several high-precision resistors wired in series and equipped with a rotary switch that alters the effective "length" of the chain to change the effective value of the resistance within the voltmeter circuit.

This technique sounds more complicated than it really is, and to prove the point, we'll design the appropriate resistor chain for the seven-range "typical" VOM voltmeter section we discussed earlier.

We start by examining our seven-position rotary selector switch: This is a simple device that consists of seven metal contacts mounted on a ring of insulating material and positioned in an arc around a moveable wiper arm. The wiper is fastened to a shaft equipped with a knob. Thus, by turning

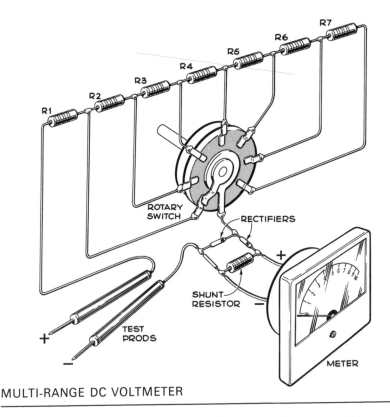

MULTI-RANGE DC VOLTMETER

the knob, the wiper can be shifted to any of seven possible positions. In each position, the wiper makes electrical contact with only one of the metal contacts.

The resistor chain is designed, as the diagrams show, so that the first link (bottom resistor of the chain) is the appropriate-value series resistor for the voltmeter's lowest-voltage range. When the switch is turned to position #1, only this resistor (which is labeled resistor R1) is in series with the microammeter and the test leads.

Resistor R2, the second link in the chain, has such a value that the sum of resistors R1 and R2—their electrical resistances in series—is the appropriate-value series resistor for the second-lowest voltage range—0-to-5 volts—of the voltmeter. With the switch at position #2, these two resistors, connected in series, are placed in series with the meter and the two test leads.

In the same way, R3 must have a value such that the sum of it plus the resistance values of R1 and R2 totals up to the correct series resistance for a 0-to-10-volt voltmeter. And, as before, turning the switch to position #3 connects R1 and R2 and R3 in series with the meter and the test leads.

Resistors R4, R5, R6, and R7 are brought into the circuit in the same fashion. Each is chosen so that the sum of its value, plus the values of the resistors that precede it in the chain, add up to the appropriate series resistance for one of the voltmeter's voltage ranges: Range 4 (0-to-50 volts); Range 5 (0-to-100 volts); Range 6 (0-to-500 volts); and Range 7 (0-to-1000 volts), respectively.

The following table gives all of the pertinent data for the resistor chain:

Range #	Value of resistor brought into the circuit	Total of preceding resistors in circuit	Total overall resistance
1 0-to-1 volt	20,000 ohms	0 ohms	20,000 ohms
2 0-to-5 volts	80,000 ohms	20,000 ohms	100,000 ohms
3 0-to-10 volts	100,000 ohms	100,000 ohms	200,000 ohms
4 0-to-50 volts	800,000 ohms	200,000 ohms	1,000,000 ohms
5 0-to-100 volts	1,000,000 ohms	1,000,000 ohms	2,000,000 ohms
6 0-to-500 volts	8,000,000 ohms	2,000,000 ohms	10,000,000 ohms
7 0-to-1000 volts	10,000,000 ohms	10,000,000 ohms	20,000,000 ohms

In the above example we used a simple seven-position rotary switch to select the appropriate resistor chain "length" for each DC voltage range. The rotary switch used inside an actual VOM is far more complex. Although it does contain a section that performs the switching function described above, it also is equipped with several other sections that enable it first to select the instruments operating mode—DC voltage, AC voltage, direct current, or resistance measuring—and then to choose a specific range within each mode.

A Circuit to Measure AC Voltage

As we have said, a moving-coil meter movement will not respond to alternating current, and so an AC voltmeter circuit that uses a moving-coil movement must incorporate some sort

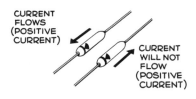

CURRENT
FLOWS
(POSITIVE
CURRENT)

CURRENT
WILL NOT
FLOW
(POSITIVE
CURRENT)

of device that converts alternating current to direct current. The device used in virtually all VOMs is the *semiconductor rectifier*. A rectifier acts like a one-way valve for electric current flow. Current can flow through the rectifier in only one direction.

If you connected a semiconductor rectifier in series with the test leads of a DC voltmeter, the meter would indeed present you with a "voltage reading" when you touch the instrument's test prods to an AC voltage source. However, you'd soon find that the DC meter's scale calibrations are worthless. The meter reading would not come close to matching the actual value of the AC voltage you are measuring. And you'd find that the meter would not respond at all to low AC voltages.

Correcting these two flaws requires a bit of circuit modification; a practical AC voltmeter is not simply a DC voltmeter in series with a rectifier. To understand why, and to understand what part the rectifier plays in the voltmeter circuit, we must first look at an alternating current in detail.

Back in 1897, the English physicist J. J. Thomson discovered the tiny particles of electricity that he called electrons. Thomson explained that each electron carries a minute electric charge, and that an electric current flow in a wire is actually the movement of electrons through the wire.

Thomson's discovery helped to explain a good many of the mysteries of nineteenth-century electrical science. An electrical conductor, for example, is basically a substance that offers little opposition to the movement of electrons, while an insulator is a substance that blocks electron flow.

A voltage source, such as a battery or generator, is therefore a device that develops an electromotive force, a force capable of pushing a stream of electrons through an insulator. The device's voltage—measured in volts—is an indication of how strong is the electron-pushing force.

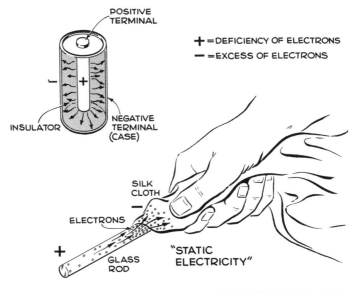

POSITIVE TERMINAL

INSULATOR

NEGATIVE TERMINAL (CASE)

+ = DEFICIENCY OF ELECTRONS
— = EXCESS OF ELECTRONS

SILK CLOTH

ELECTRONS

GLASS ROD

"STATIC ELECTRICITY"

ELECTROMOTIVE FORCE

Further, Thomson's theory explained what takes place inside a working voltage source: the device builds up a net surplus of electrons at one of its terminals, and a net deficiency of electrons at the other. For example, consider a common flashlight cell:

Every electron has a slight negative electric charge, so that the cell's negative terminal is actually a body that contains a net excess of electrons, giving it a net negative electric charge. Where did these excess electrons come from? From the positive terminal. It now has a deficiency of electrons, and so it possesses a positive charge, since a deficiency in negative charge is equivalent to a net positive charge.

This concept of surplus and deficiency of negative charge leads from the discovery that electrons are part of the atoms of all elements. This means that an electrically uncharged piece of any substance has "just the right amount" of electrons inside it. If somehow you add more electrons, you give the substance a net negative charge; if you take away electrons, you leave the substance with a net positive charge.

Moving electrons—and hence, electrical charge—around, is

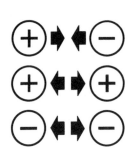

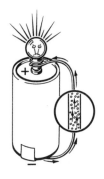

ELECTRON FLOW

not a difficult thing to do. If you stroke a glass rod with a silk cloth, friction pulls electrons off the glass, and gives it a positive charge. This kind of charge shifting is responsible for many of the phenomena we label "static electricity" effects, including the mild shock you receive if you walk across a rug on a dry day, and then reach for a doorknob.

Inside the dry cell—or any other type of cell or battery—a chemical reaction forces electrons away from the positive terminal and on to the negative terminal. And, inside a generator, the movement of a conductor through a magnetic field induces the flow of electrons through the conductor.

The pioneers of electrical science, working hundreds of years before Thomson announced his findings, had no idea what electrical charge was made of. But they did observe that a positively charged body attracts a negatively charged body, while two similarly charged bodies—either positive and positive, or negative and negative—repel each other. Since an electron carries a negative charge, it's clear that a positive body will attract electrons, while a negative body will repel electrons. Thus, if you wire a flashlight bulb across a dry cell, electrons will flow through the bulb's filament, pushed by the cell's negative terminal, and pulled by the cell's positive terminal.

Strictly speaking, electricity is the flow of electrons, and always flows away from a negative terminal and toward a positive terminal. Unfortunately, the truth tends to confuse the arithmetic of electrical calculations, since it involves negative numbers. And so, by agreement of almost everyone who works with electronics, electricity is assumed to be made of a mythical stream of positive charges that flows away from the positive terminal, toward the negative terminal. A bit of thought will verify that a stream of mythical positive charges moving to the left, through a wire, is equivalent to a stream of real negative charges moving to the right.

A *direct current* is simply an uninterrupted flow of electrons through a conductor, in one direction, much like a stream of cars along a one-way street. A dry cell or battery are the commonest sources of direct current. Connect a conductor across a dry cell, and a direct current will flow through the conductor until the cell is exhausted. A great many electrons must flow at the same time to add up to an appreciable current. For example, a direct current of 1 ampere means that 6,280,000,000,000,000,000 electrons are whizzing through the conductor each second.

An *alternating current* is a flow of electrons that changes its direction of flow periodically. Electrons first move through the conductor in one direction, and then shift gears and move in the other direction.

An alternating current flow is produced by a voltage source whose terminals repeatedly reverse their polarity. Perhaps the most common is the dynamo in your local power station. Sixty times each second its output terminals change polarity—each switching repeatedly from positive to negative to positive to negative, and so on—so that the electrons flowing in your home's power lines reverse direction sixty times each second.

In the case of a dynamo, its output voltage—the voltage measured, or seen, between its terminals—varies along with the change in terminal polarity. The output voltage varies from a positive maximum to zero, and then to a negative maximum.

The smooth curve traced by the value of the output voltage

when plotted against time is called a *sine curve*. Note that the illustrations show only a small fraction of a second; during a complete second, the voltage will vary from positive to negative and back to positive exactly 60 times.

One point to keep in mind is that an alternating voltage source doesn't necessarily have to produce the smooth output-voltage change characteristic of a dynamo. In other words, although the most familiar source of AC voltage—your local power plant—generates a sine-curve-shaped output-voltage variation, many other sources do not. The illustrations show a few voltage curves that fall under the general classification of "alternating voltages." This is an important point, since the shape of the voltage curve affects AC voltmeter accuracy. We'll say more about this later, but remember that the AC voltmeter circuit in a VOM is specifically designed to measure sine-curve-shaped waveforms, and that its scale is not calibrated for reading other types of alternating voltage.

An alternating current can also be represented by this type of diagram. The output of an AC dynamo, when plotted against time, produces a sine-wave-shaped curve, as you would expect. Here, the vertical dimension represents current at any instant of time.

One common misconception about alternating currents is that the electrons within them are each alternately speeding up, then slowing down, then reversing direction, then speeding up again—all in step with the applied alternating voltage.

Actually, electrons always move at a constant speed within a conductor. At any instant of time, an alternating voltage propels a particular number of electrons through the conductor. As the amplitude and polarity of the voltage waveform changes, so does the number and direction (respectively) of the electrons making up the current.

To illustrate the process, let's consider the current flowing through a conductor connected across a low-voltage, sine-wave-shaped, alternating-voltage source.

At the instant that the voltage curve crosses the base line (horizontal axis) the voltage across the conductor is zero, so no

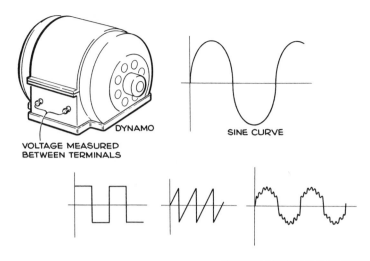

VOLTAGE MEASURED
BETWEEN TERMINALS

DYNAMO

SINE CURVE

OTHER VOLTAGE CURVES

electrons flow through it. As the curve begins to rise—representing a positive voltage increase—more and more electrons join the steadily increasing current flow, until, at the instant that the voltage curve reaches its maximum, the maximum current is flowing through the conductor. Then the voltage starts decreasing. The number of electrons flowing through the conductor decreases steadily until the voltage curve crosses the horizontal axis, at which instant the current equals zero amperes.

As the voltage source reverses polarity, electrons start flowing in the reverse direction, steadily increasing in number until the voltage waveform reaches its negative minimum. This is the point of maximum current flow in the reverse direction.

Finally, the voltage begins to rise again toward zero. Simultaneously, the number of electrons moving in the reverse direction decreases, until the voltage curve intersects the horizontal axis, and the cycle begins again. Then it would loose electrons until the current level dropped to zero, once again at the intersection of the curve with the base line.

Of course the exact number of electrons in motion at any given instant of time is proportional to the precise value of the alternating voltage propelling them through the conduc-

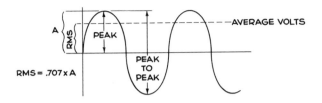

RMS = .707 x A

VARIOUS AC VOLTAGE MEASUREMENTS

tor at that particular instant.

Armed with these facts, we can tackle the question of "How do you design a voltmeter to read alternating voltages?"

A bit about averages. We are ready to design a voltmeter, but we have one more problem ahead of us: figuring out what voltage we want to measure. Strange as it seems, there are several voltages associated with alternating current.

We could measure the *peak voltage*—the "distance" between the axis and either the maximum or minimum points on the AC-voltage, sine-wave-shaped curve.

Or we could measure the *peak-to-peak* voltage—the "distance" between the maximum and minimum points on the voltage plot.

Or we could measure some sort of *average* voltage that expresses a comparison between AC and DC voltage.

Although many measuring instruments (including the VTVM, which we will discuss shortly) work with peak voltages, the AC voltmeter stage of a VOM is designed to measure a special kind of average AC voltage called the *root-mean-square* (or RMS, for short) voltage. Here's the idea:

If you connect a light bulb to an AC voltage source, it will light, just as it will if you connect it to a DC voltage source. To the light bulb, the AC and DC voltage sources can be equivalent, as far as their ability to heat its filament is concerned. Obviously, it would be convenient to say, for example, that a 110-volt AC voltage source is equivalent to a 110-volt DC voltage source, and we can, provided we deal with RMS AC voltages. In a nutshell, then, the RMS voltage is the numerical value of AC voltage that is as effective in doing work (lighting a light bulb, heating a heating coil, for exam-

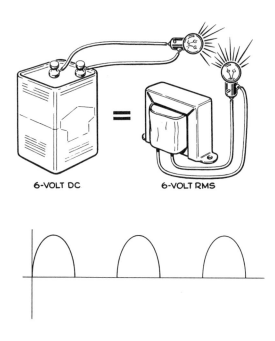

6-VOLT DC   **6-VOLT RMS**

ple) as the corresponding value DC voltage.

A 6-volt RMS AC voltage is "equivalent" to a 6-volt DC voltage, or a 120-volt RMS AC voltage is "equivalent" to a 120-volt DC voltage.

When we say "equivalent," we mean that the alternating current driven through the light bulb by a 120-volt RMS AC source does as good a job of heating the filament as the direct current driven through the filament by a 120-volt DC source.

Probably, you are used to referring to the power-line voltage in your home as 120 volts AC. Actually, you should call it 120 volts RMS, to be perfectly accurate. In effect, your local power company has set its generators so that they produce alternating current having the appropriate peak voltages so that the resulting RMS value equals 120 volts.

How do you determine the RMS value of any specific AC voltage? Although you can calculate it if you know the peak voltages, the easier way is to measure it with an AC voltmeter calibrated to read RMS voltages. This is, of course, the type of meter found in all VOMs.

The RMS voltmeter circuit. We start with a semiconductor rectifier, a current-limiting resistor, and a DC microammeter, all wired in series. When this circuit is connected across a source of alternating voltage, the source tries to propel an alternating current through the components. But it can't do the job. The rectifier (which acts like a one-way valve for electrons) only permits electrons to move in one direction through the three circuit elements. The net result is that the "bottom half" of each alternating-current cycle is clipped off, and pulsating direct current flows through the circuit. Pulsating DC is simply a flow of electrons that periodically changes in intensity, but still moves only in one direction.

The current level rises smoothly to a peak value, and then declines smoothly to zero—the rise and fall follows the shape of one-half of a sine curve. Then the current remains at zero for a short period of time, before beginning to increase again. This short-time delay is, of course, equal to the period of time required for the other half—the clipped-off half—of the AC voltage curve to trace its "bottom half."

A pulsating direct current will make the microammeter needle deflect. In effect, the inertia of the needle, and the friction of its bearings, damp out the rapidly rising and falling bursts of current. The meter literally "averages" the incoming current pulses into a steady DC reading. (*Note:* Some especially fast-acting meter movements will respond to the individual pulses: the meter needle will jitter back and forth somewhere on the scale. The meters used in VOMs, though, are specially damped to avoid this type of confusion.)

As with the DC voltmeter, the value of the series resistor determines the circuit's operating range. But, unlike our earlier example, a straightforward Ohm's Law calculation can't be used to compute the resistor's value. The reason is that a pulsating direct current is not equivalent to a nonpulsating direct current. Because its value periodically changes, the meter needle will stabilize at a significantly lower reading, compared to the reading it would indicate if fed pure direct current whose amperage is the same as the pulsating direct current's peak amperage. This means we must add a correction factor to our calculation, that equalizes the relative efficiencies (at swinging the meter needle) of pure and pulsating direct current.

And, of course, we have another correction factor to consider: the factor that matches up RMS voltage with the corresponding level of pulsating direct current it produces in our simple rectifier/resistor/meter series circuit.

The necessary arithmetic to calculate the correction factors is a bit too involved to show here. Instead, we'll simply state the end result when both factors are combined together to produce a single overall correction factor that allows us to use Ohm's Law once again to figure out the proper series resistor values.

Every AC RMS volt is approximately .318 times as effective at swinging the meter needle as is a pure DC volt. *Note:* This factor is correct only when we deal with this particular type of voltmeter circuit.

All of this means that to calculate the proper value of the series resistor to measure any particular RMS AC voltage range we mentally eliminate the diode—simply pretend it isn't there—and then design a DC voltmeter to measure .318 times the full-scale AC range we desire.

Thus, for example, if we want a 0-10-volt AC RMS voltmeter, we must go through the design procedure to build a 0-to-3.18-volt DC voltmeter. We'll do exactly this, shortly, as soon as we make two minor modifications to the circuit that will improve its low-voltage-reading accuracy.

Our job is to compensate for the fact that a semiconductor rectifier is not a perfect one-way current valve: It allows a small "leakage" current to flow backwards through itself during the negative half of each voltage wave. Unhappily, this backwards current is almost as great as the forward current we want to flow, when the circuit is measuring low AC voltages. The net result is that alternating current rather than pulsating direct current flows through the microammeter, when low AC voltages are read.

And another less important flaw is that the rectifier acts like a relatively high-value resistor when a tiny current flows through it. At higher current levels, the rectifier's internal resistance drops way down. Since this changing resistance is wired in series with the series resistor, R, it tends to throw off circuit accuracy when low AC voltages are measured.

Two additions, as shown in the illustration, cure both problems:

1. A second semiconductor rectifier wired across the meter/series resistor circuit acts to "short-circuit" the meter during the negative half of each voltage cycle. This reduces the amount of "backward" current flow that moves through the meter.

2. A shunt resistor, wired directly across the meter, effectively reduces the meter's current-reading sensitivity. We'll discuss shunts, and their operation, in the next section. For now, we will simply state that when the shunt resistance is equal in value to the meter's internal resistance, the meter's effective sensitivity is cut in half. For example, a 50-microampere meter movement becomes a 100-microampere movement.

The reason for the shunt resistor is to increase the pulsating direct current flow through the rectifier. A less-sensitive meter requires that the series resistor have a lower value (so more current can flow to deflect the meter needle). This in turn means that more current flows through the rectifier, decreasing its effective internal resistance, and improving circuit accuracy.

RMS VOLTMETER

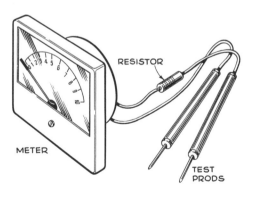

RESISTOR

METER

TEST PRODS

We start our calculations by fixing the value of the meter shunt resistor. As we stated earlier, the meter we use has an internal resistance of 300 ohms. Thus, a shunt resistance of 300 ohms will lower the meter's sensitivity to 100 microamperes full-scale deflection.

Next, we determine the value of the series resistor, R. (Note that in our calculations we can forget about the presence of the second semiconductor rectifier, as well as the first. It acts like an open circuit during the positive half of every voltage wave, the only half our circuit converts into a meter reading.)

As we've said, for a 0-to-10-volt AC RMS voltmeter, we use the same value series resistor as required for a 3.18-volt DC voltmeter:

$$100 \text{ microamperes} = \frac{3.18 \text{ volts}}{R_{\text{ohms}}}$$

$$\dots \text{thus} \dots \ R = 31,800 \text{ ohms}$$

As before, we can neglect the effect of the meter's resistance (in combination with the shunt) since it is much less than 31,800 ohms.

The addition of a series resistor chain and multiposition switch, as was used in the DC circuit discussed earlier, equips this basic AC voltmeter circuit for multi-range operation. The complete multi-range circuit is shown in the illustration. Below, we'll simply list, in tabular form, the specific values of the different resistors in the chain. They are computed in exactly the same way as were the DC voltmeter resistances.

	Range #	Value of resistor brought into the circuit	Total of preceding resistors in circuit	Total overall resistance
1	0-to-1 volt	3,180 ohms	0 ohms	3,180 ohms
2	0-to-5 volts	12,720 ohms	3,180 ohms	15,900 ohms
3	0-to-10 volts	15,900 ohms	15,900 ohms	31,800 ohms
4	0-to-50 volts	127,200 ohms	31,800 ohms	159,000 ohms
5	0-to-100 volts	159,000 ohms	159,000 ohms	318,000 ohms
6	0-to-500 volts	1,272,000 ohms	318,000 ohms	1,590,000 ohms
7	0-to-1000 volts	1,590,000 ohms	1,590,000 ohms	3,180,000 ohms

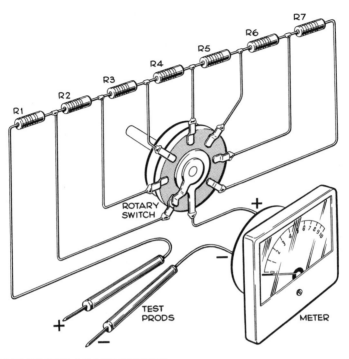

MULTI-RANGE AC VOLTMETER

A Circuit to Measure Resistance

There are several practical resistance-measuring circuits—or *ohmmeters*—in use; the one we will discuss has a voltmeter at its heart. The circuit uses a very common arrangement of two resistors wired in series, which is called a *voltage divider,* a low-voltage dry-cell battery, and a conventional DC voltmeter circuit that continuously measures the voltage across one leg of the divider. Before we consider the circuit as a whole, let's look at its component parts.

A voltage divider is simply a chain of two or more resistors wired in series across a voltage source. The simplest divider circuit—a chain of two resistors—is used in an ohmmeter. The illustration shows how this circuit works:

Consider two resistors, R1 and R2, wired in series across a

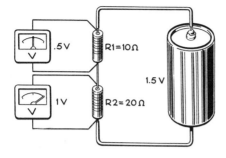

1.5-volt dry cell. A 0-to-1-volt DC voltmeter is connected across each resistor. For the resistance values shown (R1 = 10 ohms, R2 = 20 ohms) notice that the two voltmeters have different readings: VM1 reads .5 volts, VM2 reads 1 volt. Note that the sum of the two voltage readings equals 1.5 volts—the voltage of the dry cell wired across the resistor chain.

With different value resistors in the chain, the readings on the two voltmeters would be different *but their sum would always equal 1.5 volts.*

In short, a voltage divider splits up the total voltage applied across it into two or more voltages (the number of voltages is equal to the number of resistors in the chain). The relative voltage values measured across each resistor in the chain depends on the relative values of all the resistors, but their sum is always equal to the total voltage applied across the chain.

A simple formula lets us calculate the voltage across any particular resistor in the chain. For our simple two-resistor divider:

$$\text{Voltage across R1} = \frac{R1}{R1 + R2} \times V_{\text{total volts}}$$

$$\text{Voltage across R2} = \frac{R2}{R1 + R2} \times V_{\text{total volts}}$$

Thus, in our example R1 = 10 ohms, R2 = 20 ohms, and $^\text{v}$total = 1.5 volts, so that:

$$\text{Voltage across R1} = \frac{10}{10 + 20} \times 1\frac{1}{2} = \frac{1}{3} \times 1\frac{1}{2} = .5 \text{ volts}$$

$$\text{Voltage across R2} = \frac{20}{10 + 20} \times 1\frac{1}{2} = \frac{2}{3} \times 1\frac{1}{2} = 1.0 \text{ volts}$$

An ohmmeter circuit, as the illustration shows, is really a special form of the simple voltage-divider demonstration circuit we discussed above. R2 is a fixed precision resistor, and this time there is only one voltmeter; it reads the voltage across R2. R1 (shown in dotted lines) represents the unknown resistance—the resistor whose value we want to measure.

With an unknown resistor in place, the voltmeter will register some particular voltage reading. And, by working backwards with the second form of the relationship given above, we could calculate the value of R1 (the unknown resistance). Luckily, there's an easier way. Since R2 is fixed, as is the voltage output of the dry cells, and we know both values, we can draw a scale on the face of the meter that gives us the unknown resistance value directly.

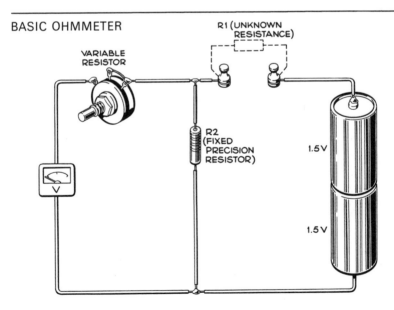

BASIC OHMMETER

VARIABLE RESISTOR

R1 (UNKNOWN RESISTANCE)

R2 (FIXED PRECISION RESISTOR)

1.5 V

1.5 V

When R1 has infinite resistance value—there is no resistor connected across the terminals—no current flows through R2, and the voltmeter reads 0 volts. On the other hand, when R1 has zero resistance—the two terminals are short-circuited together—the voltmeter reads 3 volts, since the total source voltage appears across resistor R2. Thus, 0 and 3 volts represent the "left" and "right" extremes on our meter's ohmmeter scale.

In order that the full-scale arc on the meter's face comes in to play, we must use a voltmeter that reads 0-to-3-volts DC full-scale. The simplest way of doing this is to take our earlier 0-to-1-volt DC voltmeter and add an additional resistance in series with it to bring its range up to 3 volts full scale. The illustration shows this as a 0-to-50,000-ohm variable resistor (or potentiometer). Because it is variable, this additional series resistor also serves as a "calibrating control" to help compensate for small source voltage changes as the battery of dry cells ages. We'll say more about this shortly. Right now, think of this component as a 40,000-ohm fixed resistor that combines with the voltmeter's original 20,000-ohm resistor to produce a 0-to-3-volt full scale DC voltmeter.

Now, suppose that the fixed resistor R2 has a value of 100 ohms. If the unknown resistor then has a value also of 100 ohms, the source voltage will split equally across R1 and R2, and the voltmeter will read 1.5 volts (the needle will rise halfway up the scale). Thus, midpoint on the ohmmeter scale we are drawing on the voltmeter's face corresponds to 100 ohms.

If, in another test, the unknown resistance is 1000 ohms, the voltage across R2 will be:

$$V_{R2} = \frac{100}{100 + 1000} \times 3 = \frac{1}{11} \times 3 = .27 \text{ volts}$$

Thus, the 1000-ohm mark on the ohmmeter scale will be about $\frac{1}{12}$ of the scale above 0 volts (which represents infinite ohms).

By doing this type of calculation repeatedly, for many different "unknowns," we can produce a complete ohmmeter scale.

Note that this scale is not linear; the high-resistance marks are crammed together tightly near the low end of the scale, while the low-resistance calibrations are spread out near the high end of the scale. This characteristic is a fact of ohmmeter life, and there is no solution for it.

However, it's obvious that this ohmmeter is almost useless for measuring high-value resistors. It's impossible to read the tiny differences at the low end (high-resistance end) of the scale.

The only practical way of getting around this difficulty is to provide the ohmmeter circuit with several different "fixed" resistors (several R2s) that can be switched into the circuit to vary the ohmmeter's center scale calibration. For example, if R2 is equal to 1000 ohms, the center point on the ohmmeter scale now corresponds to 1000 ohms, rather than 100 ohms as in the earlier example. And, if R2 equals 10,000 ohms, the

MULTI-RANGE OHMMETER

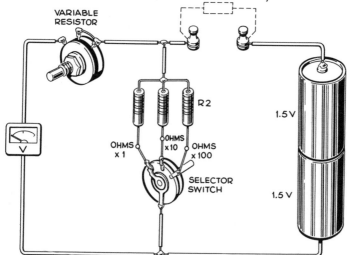

center point corresponds to an R1 value of 10,000 ohms. Note that in all these examples, the "range" of the ohmmeter circuit hasn't changed. The far left end of the scale still represents infinite ohms, and the far right of the scale still corresponds to zero ohms. It's just the center region that has been expanded for easier reading.

Note also that the same meter scale we provided for the 100-ohm fixed resistor can be used when the 1000- or 10,000-ohm resistors are brought into play. You must remember though that when the 1000-ohm resistor is in the circuit, every marking on the scale is effectively multiplied by 10; and when the 10,000-ohm resistor is switched in, every marking is multiplied by 100. Thus, we have labeled the switch positions (on the rotary switch that selects which of the three resistors will be in the circuit) in the illustration: OHMS × 1, OHMS × 10, and OHMS × 100, respectively.

One potential source of measurement error is the pair of dry cells or, rather, their tendency of developing a considerably lower output voltage as they age. A lowered source voltage means that the ohmmeter circuit will read less than full-scale deflection (0 ohms) when the two test probes are short-circuited together.

The solution to this problem is the adjustable series resistor in the voltmeter circuit that we discussed earlier. On your VOM, this control will probably be labeled "Ohms Calibrate" or "Ohms Adjust."

Simply, this control is provided so that you can quickly adjust the full-scale range of the voltmeter to exactly match the output voltage of the voltage source (pair of dry cells). You do this by first short-circuiting the ohmmeter's test probes together, and then turning the Ohms Calibrate knob to produce full-scale deflection on the meter.

This easy-to-make adjustment—which you should perform just before you take a resistance reading every time you use the ohmmeter section of your VOM, and each time you switch to a different resistance range—restores the accuracy of the ohmmeter circuit.

A Circuit to Measure Direct Current

The basic meter movement within a VOM is a direct-current measuring instrument, so that the task of the DC ammeter circuitry within a VOM is to "desensitize" the meter movement. Practically speaking, the vast majority of direct currents flowing inside electronic circuitry range from a few tenths of a milliampere (thousandth of an ampere) to one ampere. This means that a four-range instrument—0-1 ma; 0-10 ma; 0-100 ma; and 0-1000 ma (0-1 amp)—is sufficient for practical electronic use. A few VOMs include 0-to-10-amp scales. You'll rarely use this range unless you service automotive electronic gear, where currents of this magnitude are occasionally found.

BASIC DC AMMETER

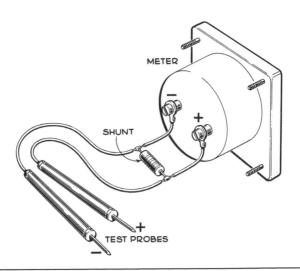

As we've said, the device that lowers the sensitivity of a DC meter movement is called a shunt. It's nothing more than a resistor placed across the movement's terminals that detours a fixed proportion of the total current flowing through the ammeter circuit (see illustration) past the meter movement.

The appropriate value of a shunt resistor can be calculated from a simple formula:

$$\underset{\text{(current through meter)}}{I_{\text{meter}}} = \underset{\text{(current through circuit)}}{I_{\text{circuit}}} \times$$

$$\underset{R_{\text{shunt}} + R_{\text{meter}}}{\dfrac{R_{\text{shunt}}}{(\text{shunt resistance})}} \dfrac{(\text{shunt resistance})}{(\text{meter resistance})}$$

The first step in using the formula is to specify the circuit's current-measuring range. Suppose we want a 0-to-100-ma DC milliammeter. This means that with a total current of 100 milliamperes ($1/10$ ampere) flowing through the circuit, we want the meter needle to deflect fully. In other words, a circuit current of 100 ma should produce a meter movement current of 50 microamperes (50/1,000,000 ampere).

We know that the meter's internal resistance is 300 ohms, so that plugging the appropriate values into the formula:

$$\frac{50}{1,000,000} = \frac{1}{10} \times \frac{R_{\text{shunt}}}{R_{\text{shunt}} + 300}$$

or

$$R_{\text{shunt}} = \frac{1500}{9995} \text{ ohms} = (\text{approx}) \ 1.5 \text{ ohms}$$

Note that since this equation is a linear relationship, a current lower than 100 milliamperes will produce a proportionately lower meter reading. Thus, we can use the linear 0-to-10 scale on the meter's face to indicate current, as long as we remember to multiply the reading by 10.

A multirange DC ammeter, like the other measuring circuits we've discussed, consists of a multiposition switch that selects among a set of shunt resistors. The circuit for a four-

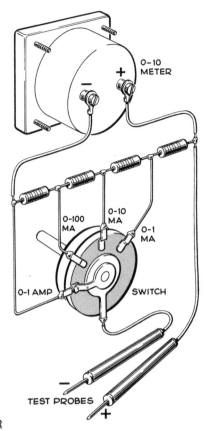

0-10 METER

0-100 MA

0-10 MA

0-1 MA

0-1 AMP

SWITCH

TEST PROBES

MULTI-RANGE DC AMMETER

range instrument is shown in the illustration, and the shunt resistance values are listed in the following table:

Range #	Value of shunt brought into circuit	Total of preceding shunts in circuit	Total overall resistance
1 0-to-1 amp	.015 ohms	0 ohms	.015 ohms
2 0-to-100 ma	.135 ohms	.015 ohms	.15 ohms
3 0-to-10 ma	1.35 ohms	.15 ohms	1.5 ohms
4 0-to-1 ma	13.5 ohms	1.5 ohms	15 ohms

As before, the linear 0-to-10 scale on the meter's face can be used for all of the above ranges, provided we always remember to multiply the reading by an appropriate "scale factor." Usually, the scale factor for each range is printed next to the selector switch (see circuit illustration).

A Word About AC Ammeters

Since you will rarely need to measure alternating current when you work with electronics, AC ammeters are not incorporated into VOM circuits. In the next chapter, we will discuss two AC ammeters that are useful if you service electrical gear. One is an adapter that plugs into a VOM, and converts it into an AC ammeter; the other is an independent instrument that uses a different kind of meter movement, and reads AC directly.

The Complete VOM

In describing the individual measuring circuits found within a VOM, we've shown a multiposition switch incorporated in each circuit. Actually, in a real VOM, these individual switches would be combined into one master rotary switch that selects the mode (what quantity the device is measuring) and the range in a single motion. The overall VOM circuit shown in the illustration (left) is made up of the circuits we've described, all tied together by such a switch.

Note that in electronics, as in other fields, there are often several ways to skin the same cat, and so the VOM you buy will probably have a different number of ranges, may have a different set of full-scale readings, and will certainly use slightly different value components in its circuitry. And it's also possible that your instrument's circuitry will be different to a degree. However, the basic operating principles described above will hold.

The Vacuum-Tube Voltmeter (Electronic Voltohm-Meter) Versus the VOM

From the outside, the vacuum-tube voltmeter—or VTVM— looks a lot like a VOM, and a closer inspection proves that the VTVM is designed to measure DC and AC voltage, and resistance—just like a VOM. Then why own both instruments?

The main reason is that a VTVM doesn't "load" a circuit when you are taking voltage measurements.

To see what this means, consider the simple voltage divider circuit illustrated on page 248. It consists of two 1,000,000-ohm resistors wired in series across a 1.5-volt dry cell. Obviously, this isn't a practical circuit, but it does resemble one type of circuit element you'll find used in many tube and transistor circuits.

Since both resistors are of equal value, the voltage divider formula tells us that one-half of the cell voltage—or ¾ volt—should "appear" across each resistor.

Unfortunately, if we measured the voltage across either resistor—say R2, the bottom resistor of the chain—with a VOM set to its 1-volt DC range, we would read far less than ¾ volt. The reason is that the VOMs voltmeter circuit *acts like a shunt,* and quite literally short-circuits R2 when the test leads are connected.

After all, the 0-to-1-volt DC voltmeter we talked about above consists of a meter movement, whose internal resistance is 300 ohms, in series with a 20,000-ohm resistor. The total resistance of the circuit is 20,300 ohms. And, 20,300 ohms "looks like" a short circuit to 1,000,000 ohms. To prove the point, plug these resistance values into the shunt resistance formula given above: (The meter, you'll remember, is acting like a shunt.)

$$I_{shunt} = I_{total} \times \frac{1,000,000}{20,300 + 1,000,000}$$

$$= I_{total} \times \frac{1}{102} \text{ (approx)}$$

This means that less than $1/100$ of the current passing through R1 gets through R2—the rest is detoured through the voltmeter circuit.

This is important, since now the circuit has, in effect, been totally changed. Resistor R2 has such little effect on the circuit, we might as well forget it is there, and consider instead the "equivalent" circuit shown in the illustration.

Note that now our circuit consists of a 1,000,000-ohm resistor in series with a 20,300-ohm resistor (the internal resistance of the meter circuit), wired across the dry cell.

The voltage divider formula shows that the voltage across the voltmeter equals:

$$V_{voltmeter} = 1.5 \times \frac{20,300}{1,000,000 + 20,300} = (approx) \frac{1}{50} \text{ volt!}$$

This is the voltage that the voltmeter will indicate, not ¾ volt DC.

As you have probably observed, the degree to which a DC voltmeter loads down a circuit it is measuring depends primarily upon the value of its series resistor, which in turn depends on the sensitivity of its meter movement. If, for example, the meter movement in our circuit had a full-scale sensitivity of 10 microamperes, rather than 50, the required value for the series resistor (to make a 0-to-1-volt DC voltmeter) would be 100,000 ohms: $R_s = 1 \div .000010 = 100,000$ ohms.

Or, if the meter was less sensitive—say it had a full-scale sensitivity of 1 milliampere—the required series resistor for a 0-to-1-volt DC voltmeter would be 1000 ohms: $R_s = 1 \div .001 = 1,000$ ohms.

Clearly, the "loading" of a circuit by a voltmeter becomes a problem only if the circuit's resistance is comparable to, or greater than, the "input resistance" of the voltmeter. We'll say more about this in later chapters.

How much a voltmeter loads a circuit it is measuring is a measure of the voltmeter's sensitivity, and for any voltmeter circuit this is specified in terms of an ohms-per-volt value. The voltmeter sections in all VOMs that use 50-microampere meters have a sensitivity of 20,000 ohms per volt. This means that there are 20,000 ohms of series resistance in series with the meter movement for every volt of full-scale deflection on any range.

As we've said, the 0-to-1-volt range has a 20,000-ohm resistor; the series resistance table given earlier proves the point

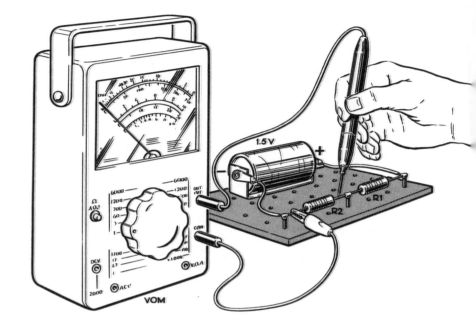

for the six other ranges. For example, the 0-to-5-volt range uses a 100,000-ohm series resistance. Thus, 1000,000 ÷ 5 = 20,000 ohms per volt. And, the 0-to-1000-volt range has a series resistance of 20,000,000 ohms, which also works out to 20,000 ohms per volt.

An inexpensive VOM, which uses a 0-to-1-milliammeter meter movement, has a voltmeter section sensitivity of only 1000 ohms-per-volt, and will thus load down circuits it measures to a much greater degree than a 20,000 ohm-per-volt instrument. Low-cost test meters of this type are therefore suitable only for measuring voltages in very low-resistance circuitry—the type of circuitry found primarily in electrical appliances and automotive electronic gear, rather than in electronic equipment.

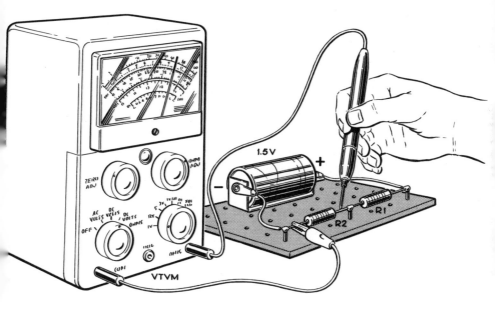

AC voltmeter sensitivity is also specified as an ohms-per-volt figure. For the circuit we designed above, the sensitivity is 3180 ohms per volt.

The VTVM presents less of a circuit-loading problem than the VOM because its internal circuitry isolates the meter movement from the circuit being measured. Simply, the basic reason a VOM loads a circuit is that electrical energy is being drawn from the circuit in order to swing the meter needle. A VTVM circuit includes an electronic amplifier, built around a vacuum tube, which drives the meter movement. Consequently, a much smaller amount of energy is drawn from a circuit during a voltage measurement—just enough to activate the amplifier circuit.

Unlike a VOM, the input resistance of a VTVM doesn't

change as the full-scale voltage range changes: the input resistance of nearly all popular VTVMs is a constant 11,000,000 ohms, on all voltage ranges.

Inside a Vacuum-Tube Voltmeter

The heart of the VTVM circuit is the "DC meter amplifier" stage. As we've said above, this stage isolates the instrument's meter movement from the circuitry or components being tested. The actual amplifier circuitry is too involved to be described here; however, it is worthwhile to consider a simple analogy that almost perfectly demonstrates how the stage functions.

Picture two 1000-ohm potentiometers, both wired in parallel across a 6-volt lantern battery. A potentiometer is basically a resistor that has been shaped into a ¾ circle, and has been equipped with a movable "wiper" arm that can be shifted, by rotating its attached shaft, to make an electrical contact anywhere along the resistor's surface. The specified resistance value of a potentiometer refers to the total resistance value of its curved resistor element. In our example, the value is 1000 ohms.

In effect, a potentiometer is a one-piece, self-contained, voltage-divider circuit. If its wiper arm is positioned exactly at the midpoint of the resistance element, a DC voltmeter, wired to the circuit, would read exactly 3 volts. The potentiometer acts like two 500-ohm resistors in series, with the wiper representing a connection to their common connection point.

For another example, look at potentiometer B. Its wiper is positioned exactly ⅓ of the resistance element's length from its "bottom" end. Thus, it acts like a 666 ⅔-ohm resistor connected in series with a 333 ⅓-ohm resistor. And a voltmeter hooked up as above will read exactly 2 volts. You can prove the point by plugging these resistance values into the voltage divider formula given earlier.

But now, suppose that a voltmeter is wired between the two potentiometer wiper arms. This voltmeter will indicate the voltage difference between the two arms.

If, for example, both arms are set at the midpoints of their respective resistance elements, the voltage of each (measured between the arms and their common bottom connection point) will be 3 volts. Thus, the voltage difference between them will be zero, and that is exactly what the voltmeter will read.

On the other hand, if wiper arm A is set ⅔ the distance from the "bottom," and wiper B is set at ⅓ the distance, the voltmeter will read 2 volts—the difference voltage, as the illustration proves.

Note that if the positions of the wiper arms were reversed (wiper A at ⅓ setting, and wiper B at the ⅔ setting) then the difference voltage between the two wipers would be -2 volts, and we'd have to reverse the voltmeter leads to get a reading.

A circuit configuration similar to this is at work inside a VTVM, only instead of two potentiometers, the VTVM uses a two-section vacuum tube, each section of which functions as an electronic voltage divider. In operation, the tubes are adjusted for perfect balance by means of a zero-set control (a potentiometer wired to one of the tube's sections) until the DC meter movement connected between them reads exactly 0 volts. Then, a DC voltage applied to the other section will unbalance the circuit, and produce a reading on the voltmeter.

In effect, the circuit acts exactly like a low-voltage-range DC voltmeter, with one important exception: The tube circuit has practically an infinite input resistance, and it presents virtually no loading to the circuit being tested.

Practically speaking, though, this circuit, by itself, is not a very useful one. The reason is that it works properly only across a relatively small voltage range—say, from zero to 3 or 4 volts DC. How about measuring higher voltages? The solution is obvious—a simple switch-selected voltage dividing chain, using a number of precision resistors, similar to the setup used in a VOM.

Of course, a voltage divider, since it is just a string of resistors wired in series, does act like a load to the circuit under test. Fortunately, though, it's possible to make this loading effect very small by making the resistance values used in the chain very large. In virtually all popular VTVMs, the total voltage-divider resistance—the sum of all the resistors in the chain—is 10 megohms (million ohms). A 1-megohm series resistor, built into the VTVMs probe, raises the instruments input resistance, on DC voltage ranges, to 11 megohms, regardless of the position of the range selector switch.

VTVMs also measure AC voltages; they are equipped with a rectifier stage, usually built around a vacuum-tube rectifier. However, unlike the rectifier circuitry used in a VOM, a VTVM rectifier stage normally produces a DC voltage output proportional to the peak, or sometimes peak-to-peak input voltage rather than RMS voltage. This difference isn't too important when you measure sine-wave-shaped AC voltages, since the direct correspondence between a sine wave's peak and RMS voltage values can be compensated for in the meter-movement scale's calibrations. Keep in mind, though, that when you read the values of non-sine-wave voltages with a VTVM, the RMS scale value can't be used; only the "peak AC voltage" scale's reading is accurate.

Another point to consider is the so-called "turnover effect." This is a characteristic of any AC rectifier circuit that responds to peak voltages. Simply stated, it's possible for the positive and negative peaks of a complicated non-sine-wave, AC waveform to have different peak values, as shown in the illustrations. Thus, reversing the VTVM's connecting leads, as shown, would produce different AC voltage readings. Clearly, though, a peak-to-peak reading rectifier stage is not susceptible to this kind of trickery, since the waveform's peak-to-peak reading "looks" the same to the VTVM regardless of which way the leads are hooked up.

Usually, the turnover effect is not a serious problem. I bring it up here just to bring it to your attention: Do not be surprised if reversing the leads on your VTVM produces a different AC voltage reading.

Contrary to their names, most VTVMs also measure resistance. As with VOMs, several different ohmmeter circuits can be used, but a typical configuration is shown in the illustrations. In it, the unknown resistor becomes a shunt across precision resistor R_b.

With no unknown resistor in place across the test leads, the "ohms adjust" control is adjusted for a full-scale meter reading. This represents infinite ohms. When the unknown resistor is connected to the circuit, the characteristics of the two-resistor voltage divider (composed of R_a and R_b) change, and the voltage reading on the meter drops. As with the VOM, the change in reading is nonlinear—as shown in the typical VTVM scale face shown in the illustrations—and the ohmmeter scale must be calibrated accordingly. If the unknown resistance is zero ohms—if it is a short circuit, in other words—the meter will read zero.

Again, as with the VOM, several resistance "ranges" can be incorporated by providing a set of different R_a's and R_b's that are switched into the circuit by the master range selector switch.

Incidentally, the function of this selector switch, as shown in the overall block diagram, is much the same as the function of the VOM rotary switch, described earlier. The range values listed next to the switch positions are typical of those found in many VTVMs, although the model you buy may be scaled somewhat differently.

The one block in the block diagram that we haven't talked about yet is the power supply. It of course supplies DC power (at moderate direct current voltages of around 150 volts) to operate the tube circuitry, and AC power (at low voltages of 6.3 volts or 12.6 volts AC) to light the tube's filament. And it is the power supply that supplies the electrical energy needed to swing the meter's needle. Once you appreciate this point, you can readily understand why a VTVM doesn't load down a circuit it is measuring. This is the essence of the concept of isolation.

The vacuum tube inside a VTVM quite literally functions as an electronic valve, which controls the flow of current through the meter movement. Only a tiny bit of electrical energy—enough to activate the "valve"—need be stolen from the circuit under test in order to produce a substantial corresponding current flow through the meter. Because of this, a VTVM can use a less-sensitive meter movement than a VOM. Typical models are equipped with 0-to-200-microampere DC movements.

Transistor Voltmeters

In recent years, transistors have replaced vacuum tubes in a myriad of applications. The VTVM is about to become one of them, thanks to a semiconductor component called the *field-effect transistor*. This solid-state device possesses many of the desirable characteristics of a vacuum tube—including a very high input resistance—and yet retains all of the familiar transistor traits: small size, cool operation, and low power requirements that can easily be met by a battery.

As you'd expect, test instruments built around these devices are called *transistor voltmeters* or TVMs, for short.

As in a VTVM, the transistor functions as an amplifier that isolates the meter movement from the circuit under test, and the operating characteristics of typical TVMs closely match those of popular VTVMs.

Which is better? I've used both, and, as this is written, I feel that the advantages of the TVM are lower weight and freedom from being tied down to the power lines, both a result of the TVM's battery-powered operation. The VTVM on the other hand costs less; at this time about half the price of a comparable TVM.

Both are equally accurate, and probably as reliable. The occasional tube replacement required by a VTVM (usually after many years) is offset by the probable vulnerability of the TVM's field-effect transistor to damage caused by accidental overloads (when you take a measurement with the instrument set to a too-low range).

Which Instrument Do You Use Where and When?

In this chapter, we've discussed four decidedly different electronic test instruments that seem to measure much the same things. A fair question at this point is: "Why should I own all four?" The best way to answer this question is for me to list the main virtues of each device. You'll quickly see that each of the instruments deserves a berth on your workshop's instrument shelf.

Continuity checker. This is probably the instrument that you'll use most often. I've found that fully 70 percent of my routine troubleshooting and assembly checks are simple continuity tests. An ohmmeter, either in a VOM or a VTVM, can do the job—zero ohms represents "continuity"—but a continuity tester is much handier, for several reasons:

It's a portable device, in every way. That's helpful when you make checks away from your workbench, as when you install or troubleshoot electronic gear in your car.

It's small enough to perch comfortably on a chassis while you're making tests, so it doesn't tie up valuable workbench space. You'll appreciate this factor most when you work on an unwieldy chassis.

You don't have to look directly at it when you are making a test. Unlike a VOM or VTVM, there's no meter to read, and you'll find that you can spot the glowing bulb easily out of the corner of your eye.

The continuity tester provides a more positive test for continuity than either an ohmmeter in a VOM or in a VTVM. The reason is that when an ohmmeter is set to a high-resistance "range," a resistance of 40 or 50 ohms produces a meter reading very close to zero ohms. This effect is accentuated if you don't set the "Ohm's Adjust" control on the VOM or VTVM properly. As we said earlier, though, a continuity checker will not light unless the circuit it's monitoring has a very low resistance. Thus, unless you adjust an ohmmeter carefully and make sure it's on a low ohms range—which is unlikely when you make a fast continuity check—a continuity checker is less likely to give you misleading test results. In short, it's a specialist at the job of checking continuity, and is therefore handier to use.

Finally, a continuity checker is such a simple device that it is virtually unbreakable—VOMs and VTVMs aren't. A carelessly caused fall can shatter their cases or switch assemblies (often made of plastic), or ruin their delicate meter movements.

Neon tester. Clearly, many of the points made above—portability, simplicity, durability, small size, etc.—apply as well to the neon high-voltage indicator. It's a perfect tool for making fast voltage checks throughout a vacuum-tube chassis, and through much of a transistorized chassis. As we shall see later, this is often a preliminary step in troubleshooting.

A distinct advantage of the neon tester is its ability to work over a wide voltage range without the need of adjusting range switches. As we've said, it will indicate the presence of any AC *or* DC voltage of between 65 and 600 volts. Of course, it can't tell you the actual voltage, in numbers—not even the most practiced eye can read much into the apparent brightness differences caused by connecting the unit to different voltages—but it can still be a great timesaver. Here's why:

Safety demands that when you take a voltage reading with a voltmeter—either a VOM or VTVM—you start by setting the voltmeter's range to a high value, greater than the highest voltage present in the chassis, *not* the range for reading the voltage you expect to find. The reason is that it is possible for a faulty component to allow this high voltage to be unexpectedly present at the terminal you are about to measure. Once, I worked on a chassis in which a faulty capacitor allowed 400 volts DC to reach a terminal where there was only supposed to be 2 volts DC. Had I measured directly with my test meter set at a low-voltage range, the instrument would certainly have been damaged or destroyed. Once you've verified a voltage reading on the high-voltage range, of course, you can simply switch down to a more appropriate range.

The single-range neon tester doesn't require any precautions of this type, and so it is the fastest-to-use instrument in your toolbox for indicating the presence of moderate-to-high voltages.

VOM or VTVM or TVM. The sections above on the VOM

and VTVM (or TVM) have outlined each instrument's capabilities, so below I will simply tabulate a few specific facts about the pair:

A VTVM does not load a circuit being measured; a VOM presents a substantially greater load at low-voltage range settings.

A VOM usually is equipped to measure DC currents; no popularly priced VTVMs have this facility.

A VTVM usually has more voltage and resistance measuring ranges than a VOM, and its low-ohm resistance ranges are capable of measuring low resistance values with much greater accuracy than the medium- to high-resistance ranges on a VOM (almost no VOMs include very low ohm ranges of the kind found on VTVMs).

A VOM is easier and faster to operate than a VTVM or TVM; the main reason is that the latter's "zero set" control (to electronically zero the meter needle) usually must be adjusted when the selector switch is moved to a new position.

A VTVM is not a truly portable instrument since it is tied to the AC power lines (a trait not shared with the TVM); the VOM is a completely portable device.

Test Instruments
You Don't Really Need

A quick glance at any electronic supply-house catalog will turn up several electronic test instruments that you really don't have to own, but probably will want to. They include.

The *oscilloscope*, the aristocrat of test instruments, which enables you to "look inside" virtually any electronic circuit and observe the flow of signals from stage to stage.

An assortment of *signal generators* for circuit testing, calibration, and adjustment.

The *signal tracer*, the electronic bloodhound that helps you pinpoint the faulty stage in a malfunctioning electronic circuit.

A selection of *component testers*, such as tube and transistor checkers, which can diagnose the ailments of electronic parts.

Special-purpose *test meters* that measure a variety of electrical and electronic quantities not within the capabilities of a VOM or VTVM. The AC ammeter, high-current DC ammeter, and low-level AC VTVM are examples.

All of these instruments, like the extras in your electronic toolbox, make useful—but not essential—additions to your test equipment shelf, and we will discuss them in this chapter.

The Oscilloscope

An oscilloscope is a device that makes the invisible visible, at least, electronically speaking, where the invisible things we are interested in seeing are the galaxy of different electronic signals present in electronic circuitry. Connect an oscilloscope to the various stages of an audio amplifier, for example, and you can follow the progress of the ever-growing, jumbled audio signal as it weaves its way from the input to the output

The oscilloscope is the most versatile electronic test instrument. It lets you *see* the shape and amplitude of signals flowing through a circuit. Professional circuit designers and technicians use this instrument regularly.

terminals. The luminous trace on the oscilloscope's screen shows you the signal's every up, down, and twist; as you move the instrument's probe from stage to stage, you can view each stage's effect on the signal.

The heart of an oscilloscope is its long, funnel-shaped cathode-ray tube, or CRT, a first cousin to the picture tube in your TV set. In the tube's neck is an electron gun which shoots a needle-thin beam of electrons—several billion each second—at the phosphor-coated face of the tube. As the electrons strike this thin coating, their energy is transferred to the phosphor molecules, making them fluoresce, or glow.

In the absence of any external forces, the beam would soar straight down the center of the tube, striking the dead center of the CRT's faceplate, and producing a single glowing spot on the phosphor-covered screen. However, the beam passes through an assembly of four deflection plates. When appropriate voltages are applied to the plates, the beam will deflect—or bend—and can be made to strike any desired point on the screen. This beam-bending process, incidentally, is simply another member in the family of electrostatic attraction phenomena that we talked about in the previous chapter. The electron beam is a stream of negatively charged particles, and so a positive voltage applied to any of the deflection plates will make the beam arc towards the plate, while a negative voltage applied to any plate will bend the beam away.

The four plates are arranged in two pairs: one positioned horizontally, the other vertically. The horizontal pair can move the beam from left to right, and the vertical pair can move the beam up and down. All it takes is a suitable deflection voltage.

For example, a positive voltage applied to plate H1 (the left horizontal deflection plate) will shift the beam off-center to the left. And a positive voltage on plate V2 (the lower vertical plate) will deflect the beam downwards. If both these voltages are applied simultaneously, the beam will strike the screen down and left of center.

An alternating voltage applied between either pair of plates—say the vertical pair—makes the plate polarities change periodically. Thus, the beam is attracted toward one plate and then the other, in turn. The point at which the beam strikes the phosphor coating shifts up and down in synchronism with the alternating voltage, painting a glowing line on the screen. And, of course, alternating voltages applied simultaneously to both pairs of plates will make the beam paint a complex pattern on the 'scopes face.

The CRT's sensitivity—the voltage required to shift the beam some given distance—depends on the tube's internal construction. However, in typical tubes, a voltage of about 50 volts DC, applied between a pair of plates, will move the beam about 1 inch off-center on the screen. Because this sensitivity is too low for most applications, oscilloscopes include amplifiers that boost a signal's voltage level before applying it to the deflection plates.

The most important use of an oscilloscope is to display the voltage variation of some electronic signal with respect to time. To put it another way, the instrument creates, electronically, the same kind of voltage-versus-time plots we used to explain alternating current in the previous chapter. To do this, the signal to be plotted is fed to the vertical amplifier (which is connected to the vertical deflection plates) at the same time as a linear sweep voltage is applied to the horizontal amplifier terminals.

This sweep voltage is a sawtooth-shaped signal—produced by the oscilloscope's internal sweep generator circuit—that rises smoothly from a negative voltage minimum to a positive peak, and then drops back quickly to the minimum. With no signal applied to the vertical amplifier, the sweep voltage will draw the beam smoothly and steadily from the far left edge of the screen to the far right edge, painting a horizonal line through the center of the screen, and then—almost instantaneously—it will snap the beam back to the far left. An important point to observe is that because the sweep voltage increases linearly with respect to time, the beam moves horizontally across the screen linearly with respect to time.

When a signal voltage is applied to the vertical amplifier, *and* the sweep waveform is applied to the horizontal amplifier, the beam traces out the signal's amplitude-versus-time plot on the 'scope's screen. For example, suppose the signal is the output of a low-voltage filament transformer connected to the AC line: a low-voltage sine-wave-shaped waveform whose frequency is 60 Hz.

Clearly, the frequency of the sweep waveform—the number of times per second that the sweep voltage rises and falls—controls the trace you see. If the sweep frequency is *lower* than the signal frequency, only part of the signal waveform will be traced on the screen before the sweep voltage returns the beam to the far left edge of the screen. And, if the sweep frequency is *much higher* than the signal frequency, several cycles of the signal waveform will be painted on the screen, all crammed together. This is why all oscilloscopes include sweep frequency adjustments that allow the user to tailor the sweep frequency to the frequency of the signal he wishes to observe.

Everything we've said above, of course, applies when a nonsine-wave-shaped signal is fed to the vertical amplifier. The electron beam will trace any shape of waveform on the screen.

A practical oscilloscope includes some circuits we haven't mentioned above:

The *synchronizing*, or sync, circuitry automatically adjusts the starting point of the sweep waveform to lock it in step with the incoming signal waveform. This helps keep the visible trace steady, and jitter free.

The input *attenuator* is a "volume" control for the signal, and controls the trace's vertical size on the screen.

The *blanking circuit* automatically turns off the electron beam for an instant at the end of each horizontal sweep so that the *retrace*—the brief right to left sweep as the beam returns to its leftward starting point—is not visible.

The *power supply* provides the necessary voltages and currents to run the various circuit stages, and the high DC voltage—of between 1000 and 2000 volts—to produce the electron beam inside the CRT.

Centering controls. These adjust DC voltages applied to the deflection plates to insure that the trace is centered on the screen.

Intensity control. This adjusts the DC voltage on a control grid within the electron gun that controls the number of electrons in the beam, and hence controls the brightness of the trace on the screen. Many 'scopes also have focus controls that vary the DC voltage on electrodes within the CRT that control the shape of the beam, and hence control the sharpness of the luminous spot that is painting the trace on the screen.

What can you do with an oscilloscope? Because the instrument is so versatile it's hard to find a point to begin talking about it. I've used my 'scope to help me design circuitry. I adjust component values on a breadboard circuit until I see the desired waveform on the screen.

I've used it to help me adjust variable components in my circuits, and to calibrate other circuitry. Once again, I watch the waveform on the screen and make my adjustments until I see the waveshape I'm looking for.

I've used my 'scope to troubleshoot faulty circuitry. Here, I know what the waveforms present at different circuit stages should look like, and an out-of-shape waveform can lead me quickly to the errant stage.

You get the idea. Throughout this chapter, we'll use the oscilloscope to view the waveforms present in, and produced by, other test gear. And, throughout this book, the oscilloscope will help us visualize how components and important circuitry operate.

Signal Generators

Signals are what electronics is all about—at least 90 percent of the time. Tiny audio-frequency signals from a stereo cartridge flow into an audio amplifier, where their power is boosted to sufficient levels to move the cones of a pair of loudspeakers; minute radio-frequency signals, picked out of the air by an antenna, are transformed into video signals that control your TV set's picture tube, and audio signals that drive the loudspeaker, by the set's circuitry; repetitive signals from your car's distributor trigger its electronic ignition system.

It's not surprising then, that a group of test instruments called signal generators have been developed over the years. Depending on type, they produce output signals of varying amplitude, frequency, and wave shape—all intended for some specific adjustment, measurement, or troubleshooting application.

Signal generators are usually classified according to the shape and frequency of their output waveforms. Occasionally, also, you will find a generator named in honor of its intended application. A "dot generator," which produces a dot pattern

Audio-signal generator produces pure, audio-frequency, sine-wave-shaped signals that can be used to test and calibrate audio circuitry.

on a color TV screen used to make "beam convergence" adjustments (remove color fringing) when the set is installed, is an example.

Audio-frequency generators are designed primarily to cover the audible frequency spectrum of from 20-to-20,000 Hz, although a few available units cover higher frequencies—up to 1 or 2 megahertz—to permit their use with ultrasonic audio gear. For audio testing, the two most practical test-signal waveshapes are the sine wave (a pure audio tone at some specific frequency) and the square wave.

Sine waves are most often used to gauge audio equipment performance. For example, frequency response checks of an amplifier are made by feeding a constant amplitude signal whose frequency is slowly varied across the audio spectrum. The variation—or lack of variation—of the output signal from the amplifier, as measured by a voltmeter or oscilloscope, can

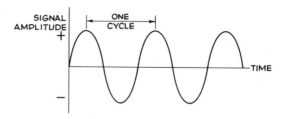

SINE WAVE

be transformed into a frequency response specification.

Distortion tests are made by feeding pure sine wave signals, at some selected frequency, into the amplifier, and then comparing the output waveform with the input waveform. If the amplifier is distortionless, the output signal will look just like the input sine wave, only larger; if the amplifier tends to distort signals passing through it, the shape as well as the size will be changed. How much? Complex test gear called *distortion analyzers* strip away the amplified replica of the original input signal, and measure the remaining distortion products to come up with a numerical distortion specification.

The square wave is a particularly interesting signal to work with since, electronically, it can be thought of as the sum of an infinite number of sine waves that have been superimposed one atop the other. To see why, we must define the term "harmonic." A harmonic of a specific sine wave is another sine wave whose frequency is an integral multiple of the original wave's frequency. For example, suppose we have a 100-Hz sine wave. Its second harmonic is a 200-Hz wave; its third harmonic is a 300-Hz wave; its fourth harmonic is a 400-Hz wave; and so forth.

Fortunately, it's not necessary to combine an infinite number of sine-wave generators to make a square-wave generator. One common way of making square waves electronically is to distort the shape of a sine wave of the desired frequency. A relatively simple circuit will flatten the sine wave's negative and positive peaks, and straighten its edges.

Sine waves, incidentally, are produced by several members of a family of circuits known as *oscillators*. We'll discuss these

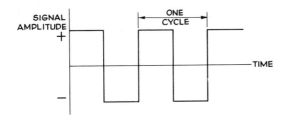

SIGNAL AMPLITUDE

ONE CYCLE

TIME

circuits in a later chapter. The oscillators used in audio-frequency generators are usually designed to produce very low-distortion sine waves.

Square-wave testing of audio equipment is a particularly powerful technique because of the multi-frequency nature of square waves—especially when an oscilloscope is used to monitor the output signal's waveform.

If an amplifier has poor low-frequency response, a low-frequency square wave—say 50 Hz—will have much of its fundamental frequency component stripped away as it passes through the amplifier. The output waveform will look droopy.

If the amplifier has limited high-frequency response, a moderate-frequency square wave—say 5000 Hz—will have most of its high-frequency components stripped away, leaving it round shouldered at the output.

If the amplifier has any built-in resonances—in other words, if it tends to exaggerate a narrow band of frequencies—a moderate-frequency square wave (say 5000 Hz) will have some of its components amplified out of proportion to the others. The result will be an output wave that displays "ringing."

Because of the usefulness of both sine and square waves in audio testing, many audio-frequency signal generators produce both types of waveforms. *Radio-frequency signal generators* cover the radio-frequency, or r-f, spectrum which begins at about 100 kHz and (practically speaking) ends at several hundred megahertz. They are all really miniature radio or television transmitters that generate imitation signals of the type normally received by radios and TV sets. No one generator

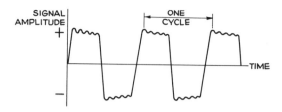

"RINGING" SQUARE WAVE

can, by itself, produce the wide variety of signals needed to test and adjust the many different kinds of r-f equipment you have in your home—including color and black-and-white TV sets, AM and FM radios, and communications gear (CB or "Ham" radios)—so that several distinct types of r-f generators have evolved.

The r-f oscillator produces a radio-frequency signal at some desired frequency. Usually, the oscillator's operating frequency is adjustable over a fairly wide range, although, as we will see shortly, some generators are designed for operation on only one or two precisely controlled frequencies. Most r-f oscillators used in generators produce output signals of roughly sine-wave shape. The waveforms are considerably more distorted than the output of audio-frequency sine-wave oscillators, but this distortion isn't important when testing r-f equipment.

The modulator. Here, we use the term in its most basic sense: a circuit that modifies some characteristic of the r-f signal (its amplitude, frequency, or waveform) in order to make it carry some other signal. For example, the modulator in an r-f generator designed to service AM radios, *amplitude modulates* (we'll explain this term shortly) the oscillator output in order to impress an audio tone onto it. Thus, when the generator is connected to an AM radio, the tone can be heard from the set's loudspeaker.

The output attenuator. This stage serves as a volume control for the generated signal, and allows the user to tailor output-signal amplitude to match the input-signal requirements of the device he is testing.

The power supply provides the required voltages and

Radio-frequency signal generator is a basic tool when you work with r-f circuitry. It produces r-f test and calibrations signals across a wide frequency range, typically from approximately 100 KHz to over 200 MHz.

currents to operate the other three stages.

The most common r-f signal generator (and the instrument you will see before you if you look up "r-f signal generators" in an electronics equipment catalog) is the instrument designed to service AM radios, and to produce r-f signals of varying frequency for general laboratory use. Typical models cover a frequency range of from 100 kHz to about 50 mHz.

The r-f "carrier" signal (at any desired frequency) is mixed together with an audio-frequency signal to produce the composite "AM" signal. Note that the frequency of this AM signal is the same as the carrier signal's frequency, but that its instantaneous amplitude is controlled by the ups and downs of the audio signal. For simplicity, the audio signal used in an r-f generator is a pure tone (usually of a frequency between 400-and-1000 Hz). Keep in mind, though, that any audio signal will work, including the complex signals that represent music and voice—the type of signals impressed on an r-f carrier at an AM radio station.

All AM radios are simply decoders for this type of signal;

inside them, a *demodulator* circuit strips away the r-f signal, leaving only the varying audio signal which is amplified and reproduced by the set's loudspeaker.

In recent years, the r-f signal generator (the type described above) has lost much of its importance to the electronic hobbyist. The main reasons are that few people build AM radios from scratch these days, and that the introduction of miniaturized solid-state circuitry has complicated radio service to the point that it is really a job for experienced technicians. And of course, most kit manufacturers design their AM radio circuitry so that the adjustable components in the set's front end and intermediate frequency (or i-f) stages can be aligned without an r-f generator. Actual on-the-air signals are used as test signals.

There are several other types of r-f signal generators that are more likely to be found on a technician's workbench than on your equipment shelf:

R-f "sweep" generator. The r-f output signal of this generator changes in frequency many times each second (usually between 50 and 100). The signal frequency smoothly sweeps from a low value to a high value and then back again, and keeps repeating.' The sweep width—the frequency difference between the low- and high-frequency end points—is usually a few megahertz.

AM MODULATION

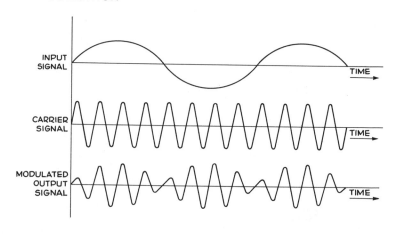

The most important use of a sweep generator is to align the i-f stages in FM and TV sets. These circuits are designed to pass a wide band of signals uniformly, and by "sweeping" across the band with a sweep generator (and simultaneously viewing the circuit's output signal on an oscilloscope), the TV technician can verify that the circuit has the required frequency response.

Marker generator. This is a high-precision r-f generator, whose output frequency is usually crystal-controlled, which generates a characteristic marker signal. It's output signal is normally combined with the output of a sweep generator, where it serves to "mark" selected key frequencies, by its remaining at constant frequency while the sweep generator output slides up and down.

Color bar and dot generator. A miniature TV transmitter that generates several patterns (most units produce more than just color bars and dots, contrary to the name) used to adjust color TV sets. The output frequency is crystal-controlled to lock it on one of the low, unused TV channels (usually either 3 or 4). A typical generator's modulator creates a series of imitation TV signals that produce the various patterns shown:

The Signal Tracer

Signal tracing is an especially powerful troubleshooting technique that is all the more useful because of its great simplicity. To put it another way, signal tracing is the most obvious way of finding out what is wrong with a complicated circuit, and it is usually a technique that works well.

Here's the idea: An electronic circuit is a signal transporting device. Electronic signals go in one end (the input) and eventually come out of the other end (the output). During their journey, as we have said, they may be transformed in some way, and the output signals may no longer look like the input signals. This happens inside a TV set: R-f signals go in, and picture and sound signals come out. But the important point is that a multistaged electronic circuit usually represents a continuous signal path—sort of like a pipeline.

If some component in an electronic circuit fails, it causes a pipeline "leak"—the signal no longer reaches the output. Finding the bad component—and patching the signal pipeline—is the essence of the process called troubleshooting.

A signal tracer is an instrument that indicates the presence of an electronic signal; and signal tracing is the technique of looking for the signal at various points along the signal pipeline. Usually, we start at the circuit's beginning, its input stage.

A suitable signal—perhaps from a signal generator—is fed into the circuit's input terminals. Then, the probe of the signal tracer is touched to the output terminals of the various stages making up the circuit, starting with the input stage, and working steadily along until we reach a stage that has no output (or has an incorrect output). Clearly, this is the stage that includes the faulty component. Once we've located it, a few voltage or resistance tests (with a VOM or VTVM), or a fast check with an appropriate component tester, will usually pinpoint the faulty part quickly.

To illustrate, let's troubleshoot a faulty monophonic hi-fi audio amplifier. Its four signal-carrying stages are shown in the block diagram (the power supply is not a signal-carrying stage; it supplies the necessary voltages and currents to run the other circuit elements).

We begin by feeding a low-level audio signal—a sine wave is fine—into the input. A loudspeaker connected to the output stage (the final bank of amplifier circuits that make the signal strong enough to drive a loudspeaker) is silent, so we know that the signal is "leaking" out of the "pipeline" through some faulty component.

Step 1. We touch the signal tracer's probe to the output of the *preamplifier* stage (it boosts the strength of the small signal fed into the amplifier's input terminal). The signal tracer indicates a moderate-level audio signal, verifying that this stage is operating properly.

Step 2. We test the output of the *tone/volume control* stage (it varies the input signal's tone and level according to the settings of the amplifier's front panel controls). Once again the signal tracer shows that the tone is reaching this point.

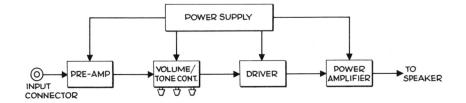

SIGNAL-CARRYING STAGES OF AUDIO AMPLIFIER

We now know that the first two stages are okay.

Step 3. We try the output of the *driver* stage (it boosts signal strength to the levels required to "drive" the output stage). The tracer indicates *no* signal. We've found the misbehaving stage. Now, our other troubleshooting techniques can take over and locate the particular component causing the trouble.

Several electronic test instruments, including the oscilloscope, and the AC voltage ranges on a VTVM, can be used as signal tracers. However, the instrument usually called a "signal tracer" is a modified audio amplifier. Its circuitry includes a loudspeaker, so that traced audio signals can be heard; a visual signal indicator (often an electronic "eye" tube similiar to the type used as level indicators on tape recorders); and a demodulator stage so that the device can be used to trace r-f signals.

A recent development is the pocket-sized signal tracer that's no larger than a pen-cell flashlight. Instead of a loudspeaker, it is equipped with a miniature earphone that you wear when you trace a circuit.

The other side of the signal tracing coin is called *signal injecting*. Here, a test signal is injected—or fed—into the input terminals of the various stages of an electronic circuit (usually starting with the last stage and working forward). For example, to test a superheterodyne radio, a test signal is first injected into the output audio stage, then into the detector (or demodulator) stage, then into the intermediate frequency (or i.f.) stage, and finally into the mixer (or front-end) stage.

The idea is the same as signal tracing. The faulty stage will not pass the signal, and so the test signal will not appear at the circuit's output when it is fed into the misbehaving stage.

The signal produced by a signal injector is usually square or sawtooth-wave-shaped audio tone that is rich in high-frequency harmonics. Thus, the same signal can be used to test audio and r-f circuitry.

Typical signal injectors are pocket-sized, battery-powered instruments. Their circuitry is simple, and they can be built into cases about the same size and shape as the probe for a VTVM. One reason for this is that they require no output indicator, as does a signal tracer: The device under test normally includes its own "output indicator." In the example above, the radio's built-in loudspeaker sounds off when the test signal enters a good stage.

Component Testers

Components are the building blocks of electronic circuitry, and making certain that the various types of components are functioning properly is the task of a specialized group of electronic test instruments called component testers.

We've already discussed one type of component tester, the ohmmeter, an instrument we can use to "test" resistors. Before we tackle the other important component checkers, though, we must first discuss the meaning—in electronics—of the terms "good" and "bad."

Probably, the component tester you are most familiar with is the simple tubechecker in your local drugstore or supermarket. Its multicolored meter tells you bluntly whether a tube is "good," or "bad," or "weak." Unfortunately, the operating state of a complicated electronic component—such as a vacuum tube—can't always be filed in one of three clean-cut categories. A tube that tests "bad" will often work perfectly well in certain types of circuits, and a tube that tests "good" may be capable of ruining the performance of another kind of circuit. This is possible because the range between "good" and "bad" encompasses a wide gray area of "may be good." It's up to the man behind the test equipment dial to interpret the results. Here's how I classify the operating condition of electronic components:

1. *Perfect.* The component meets all its specifications. Within the stated tolerance. It will function perfectly in any circuit it is designed into.

2. *Marginally good.* The component doesn't meet its specification, but will still function if this is taken into account. For example, a resistor labeled 1000 ohms (10 percent tolerance) has an actual measured resistance of 1500 ohms. Clearly, this is not a "good" 1000-ohm resistor, but it will work well in a circuit that calls for a 1500-ohm resistor. The same sort of situation often occurs with tubes and transistors. That is why a "bad" reading often doesn't tell the whole story.

Incidentally, a marginally good condition may develop as a component "ages" in a circuit. Sometimes, a slight change in component characteristic can throw a circuit out of whack. Thus, a nominally "good" test reading may hide a minor specification change that is fouling up circuit operation.

3. *Non-obvious failure.* This is simply a component failure that is so subtle that the test instruments you have on hand won't detect it. We'll say more about locating these frustrating hidden flaws in the chapter on troubleshooting; as a rule, though, the way to find them is to replace suspected components one-at-a-time until the faulty circuit works properly.

4. *Catastrophic failure.* The component is totally unserviceable; your component testers agree.

The several kinds of component testers available have one common aim: They test electronic parts by simulating—to a greater or lesser degree, depending on tester complexity—the operating conditions found in typical circuits. Thus, to understand component tester operation, you must first understand how the various components themselves work, and what they are designed to do in a circuit.

Tube testers range in complexity (and price) from simple filament testers, that only verify that the filament will conduct electricity, to emission-type testers, which measure the direct current flow through the tube from its cathode to the other elements, to mutual conductance-type testers, which actually measure the tube's amplifying characteristics.

The filament tester is the least reliable, since tubes, unlike light bulbs, can fail for many other reasons than burned-out filaments. In fact, this kind of failure is rare with modern tubes.

The emission-type tester is a better judge of tube condition. This is the kind of tester you'll find at a drug store. Its internal switching circuitry short-circuits the tube's various grids to the plate electrode, turning the tube into an almost vacuum rectifier. A DC milliammeter—the panel meter—measures the cathode current flow.

A drop in cathode emission—and hence in measured current flow—is a symptom of tube old-age. Obviously, the emission tester is also a filament checker, since a burned-out filament will not heat the tube's cathode, and no current will flow.

Conductance-type tube testers are the most reliable, but since they are also the costliest and the most difficult to use, they are rarely found in home workshops. They test a tube by duplicating actual amplifier circuitry—the tester's various knobs and dials can be adjusted to provide correct plate and grid voltages—that makes the tube function as it will when installed in a piece of operating equipment. Tube deterioration due to any cause is then readily observable on the instrument's panel meter, which reports back the tube's measured amplifying characteristics.

Transistor testers are also available in a wide range of complexity, all the way up to laboratory-grade transistor analyzers that cost thousands of dollars and monitor a test sample's every characteristic.

The most practical workshop-caliber units, though, are simpler devices that measure a transistor's DC current gain and junction "leakage." Their circuitry is very similar to fundamental transistor circuit configurations. Virtually all low-cost transistor testers also test semiconductor diodes, since the required testing circuitry is almost identical.

Battery testers are basically very low sensitivity voltmeters. They are designed to load the battery as they measure its voltage. This is necessary since it only makes sense to test a battery, by measuring its voltage, when the battery is actually at work. Thus, all battery testers incorporate switch selected "load resistances" that are connected across the battery during a test. A "good" battery will maintain its nominally rated output voltage (within about 20 percent) while it delivers a normal load current (usually specified by the manufacturer) to the resistance. The tester's switching circuit is normally arranged so that the appropriate load resistor is switched in place when the user selects either a specific voltage range (on some testers) or sets the selector switch for a particular battery type (on other units).

Troubleshooting

Sooner or later, you'll build a project that won't work properly when it is finished—I can virtually guarantee it. I base my certainty on the almost total interdependence of all the components, interconnections, and solder joints making up any electronic circuit. It's a rarity for any project to function perfectly. Usually, a bad component, a wiring error, or a cold-soldered joint will totally knock out circuit operation. And it is only good sense to admit that sooner or later you'll make an error or install a bad part.

But don't let my prediction dampen your enthusiasm for electronics. Remember that professional engineers and technicians must live with the same facts of circuit life when they build prototype circuits—it's the nature of electronic equipment.

You should, however, develop the necessary skills to find and eliminate circuit trouble-spots as they crop up. The searching process is called troubleshooting, and it is much the same whether you are trying to make a brand-new project work properly, or are trying to repair an older circuit that has suddenly stopped working. Because this book is all about building projects, we will especially emphasize those trouble-shooting techniques that will help you debug new circuitry. You'll readily recognize the methods that will also help you repair in-service circuit failures.

It goes without saying that you can only debug a valid circuit. The troubleshooting techniques in this chapter are not intended to point out circuit configuration errors. Thus, an important supplementary troubleshooting procedure is to make sure that the circuit diagram and parts list you're working from are technically correct, and are free of drawing and/or typographical errors. Practically speaking, this means sending an inquiry to the editorial office of the magazine or book that published the diagram, if other troubleshooting methods don't locate any faults.

Troubleshooting an AC powered circuit is potentially dangerous since you may be forced to poke around inside a chassis when the power is turned on. Exercise caution at all times. Connect instrument leads before you flip the power switch, whenever possible, and always unplug the line cord before you touch any terminals or lugs with tools or soldering iron.

The safety of component parts is another concern. Remember that the circuit fault may overload circuit components as well as ruin performance. Thus, try to keep the power turned off as much as possible when the device is not working correctly. In short, when you troubleshoot, make all power-on tests as short as possible. This goes for battery-powered gear, too.

Causes of Malfunction

The myriad possible causes of trouble, in even the simplest circuit, mean that you must be methodical and organized when you troubleshoot. A scattergun approach wastes time and builds frustration; a well-planned search is bound to find the trouble spot eventually and, in the process, will teach you a lot about the circuit's operation.

As a rule of thumb, trouble can be blamed on one of three causes:

1. Things you've almost done right.
2. Things you've really done wrong.
3. Bad components.

The last two items are self-explanatory; the first is worth a few words of comment since it accounts for most of the common project troubles, and also those that are the hardest to find, unless you know how to look.

A cold-soldered joint is something you've done almost right; so is a battery that is improperly mounted in a holder so that its end terminals aren't making contact; and so is a missing interconnection—the wiring would be right, except for the wire you accidentally left out.

These mistakes are tricky to spot because they often don't look wrong, like obvious short circuits or clear-cut wiring errors. And, when you turn to test equipment for help, almost/right errors may cause misleading conclusions—they often make you suspect that good components are bad. An improperly mounted dry cell, for example, could make you think that the power switch was broken, or the dry cell itself was worn out, it you took voltmeter readings.

The point is, that the first step in troubleshooting a newly finished project is to make sure that you have really finished the project. Assume for a start that all components are good, and that you haven't done anything really wrong. Then check off the items on the following list as you carry them out:

1. Inspect all plug-in and clip-in components (tubes, transistors, fuses, push-in terminals, batteries, etc.) to determine that all terminal connections are sound, and that all leads are firmly pushed into their mating socket connections. Rub away visible terminal corrosion with a hard-rubber ink eraser.

2. Examine the project for silly goofs: an off-on switch mounted upside down so that "on" really turns the circuit "off"; a multi-pin transistor socket wired incorrectly so that one or more transistor leads are plugged into unconnected socket pins; batteries installed backwards in a holder; no fuse installed in a panel-mounted fuse holder; tubes or transistors plugged into the wrong socket locations. Double-check the wiring around components you installed last. Wiring in haste breeds silly goofs.

3. Examine every solder joint carefully. Remelt any joint that looks dull, granular, or simply not perfect.

4. Jiggle the wires leading into every soldered connection. Remelt any joint that contains loose or shaky wires. Be sure to add more solder and flux.

5. Examine all chassis ground connection points. Test with a continuity checker to make sure each grounded lug or terminal makes intimate electrical contact with the chassis.

6. Look for accidental short circuits between adjacent terminals and between closely spaced component leads (especially between transistor leads, between tube socket terminals, and between switch and potentiometer lugs). These may be

caused by bent wires and/or terminals, by the protruding ends of connecting wires that you've neglected to trim off, or by loose bits of wire or solder that have lodged in place. If you are working with a printed circuit board, look for solder bridges between adjacent printed conductors. Also, flex the board slightly, and look for tiny breaks in the foil.

7. Check for missing interconnection wires. The most likely candidates are:

Wires interconnecting subchassis with panel mounted components.

Wires joining various subchassis.

Ground connections between subchassis and chassis.

Power leads between power supply, power switches, and subchassis.

Before you begin the tedious job of comparing actual interconnections with the schematic diagram, make a fast visual check. Search for possible giveaways, such as ungrounded terminal lugs holding only one wire, unsoldered ground and/or terminal lugs, and unused switch and/or control lugs.

8. If lead length and/or lead placement are known to be critical factors, try shortening and/or repositioning suspect leads (usually those leads that connect to the circuit's input stages). This is a common cure for amplifier oscillation.

Visual Troubleshooting

Putting the right thing in the wrong place is the most common example of "things you've really done wrong." Happily, a careful visual check usually reveals the error; unhappily, this kind of mistake may ruin the component(s) involved. Thus, whenever you find a right thing/wrong place error involving either semiconductor components or electrolytic capacitors, it's a good idea to remove the part(s) and test them before you reinstall it (them) correctly.

When the preliminary troubleshooting steps (outlined above) don't solve the problem, begin a point-by-point and part-by-part visual inspection of the chassis. Here are the items to cover:

Active components. Verify that transistors are correctly oriented in their sockets, or that their leads are soldered to the proper terminal lugs. Transistor base lead diagrams are normally included as part of project schematic diagrams.

Similarly, verify that tube socket wiring is correct. Many types of tube have unused pins, and a common error is to wire the corresponding blank terminals on the socket base instead of adjacent live terminals. Another frequent mistake is to wire a socket backwards; this can occur easily if the pin terminal numbers are not molded into the socket body.

Polarized passive components. "Backwards" diodes and electrolytic capacitors will often produce strange symptoms. In many cases, the circuit will seem to operate properly—but with greatly reduced performance. Curiously, experienced hobbyists tend to make polarization errors as often as do novices. Use the circuit parts list as the check list when you inspect the chassis, and verify the polarity of every polarized device.

Major component leads. These are usually identified by either geometry (how they are placed on the component's body) or by color code. Both methods can be confusing, and cause errors.

With geometric coding, the most common error is to "look" at the component in a different orientation than the coding diagram does. This mistake swaps the position of various leads, and, of course, causes a miswired lead arrangement.

Color-coded leads (as on transformers) are often confused because of the visual similarity of the colors used: dark brown and black; orange and red; blue and green; striped black and yellow, and striped black and tan.

Note: Double-check the schematic diagram of the component supplied with the part, against the component's diagram in the project's schematic. Don't simply take the project diagram's word for the meaning of the color code. It is possible that the part used by the designer of the project has a different color code than the part you purchased, expecially if you bought an equivalent component produced by a different manufacturer.

Switch and control lugs. The possibility for error here is enormous, especially when you work with complicated rotary-type switches. The best way to check for a mistake is to carefully study the contact layout of any switches, and the lug arrangement of any potentiometers, and draw yourself a simple sketch. Compare the sketch with the schematic diagram, and indicate on the sketch which leads should connect with which contacts and lugs. Then, compare the annotated sketch with the actual switch and control wiring.

Interconnecting wires. When several interconnecting wires are joined to a group of adjacent terminals (say on a tube socket or a terminal strip) mixups are hard to avoid; so is spotting any errors. The best approach is to make a copy of the schematic diagram and trace out each interconnection on the diagram as you verify the path of the corresponding wire in the chassis.

Resistors (and other color coded minor components). The visual difference between a 220-ohm resistor and a 2200-ohm resistor is the color difference (brown versus red) of a narrow band. The electrical difference is almost 2000 ohms. By its logically constructed nature, the resistor color code is full of potential traps like the above—several thousand, approximately. This is why we emphasized the importance of good illumination at the wiring table in an earlier chapter. It is all too easy to accidentally swap similar-appearing resistors. To ferret out these errors, compare the actual and diagram locations of every resistor *twice*. First, simply run through the parts list, noting the correct and actual location of each resistor. Second (as a check), make up a separate list of all similar first-digit pair resistors (resistors that have the same first two numbers in their values), and double-check that you haven't swapped any "kissin' cousins."

Labeled components. After you've gained a bit of electronic expertise, you couldn't possibly swap a 0.1-mfd capacitor with a 1.0-mfd unit. In the beginning . . . well, it's not only possible, it's probable. Follow the same procedure outlined above for resistors whenever a project includes "similar" labeled components.

Troubleshooting with Instruments

The four basic troubleshooting test instruments are the *continuity checker*, the *neon voltage indicator*, the *volt-ohm-milliammeter* (or VOM), and the *vacuum tube* or *transistor voltmeter* (VTVM or TVM). Their design and function have been discussed in Chapter 8. Think of these instruments as a team. Plan to use them together when you troubleshoot, since their detective abilities complement one another.

Broadly speaking, almost all basic troubleshooting tests fall into three categories: simple continuity checks (to establish the continuity of a coil, for example); voltage measurements that both establish the presence of a voltage and determine its magnitude (such as monitoring the voltage level between a key terminal and ground, for example); and resistance checks (to spot a "leaky" capacitor, for example).

Current measurements, made with an ammeter, aren't normally part of most chassis tests. The reasons are that installing an ammeter, so you can take a current reading, requires that you cut into the circuit. And most circuit currents can be measured indirectly if necessary by measuring the voltage developed across various resistors in the circuit, and then using Ohm's Law.

Troubleshooting with test instruments only makes sense if you have reliable knowledge of the "correct" measurement values throughout the chassis you are working on. To put it another way, unless you know what voltage and resistance values are found throughout a normally operating chassis, you can't possibly be sure—except for a few cases we'll talk about later—that the readings you make when you troubleshoot indicate a circuit flaw.

Practically speaking, there are several ways to obtain these values—sometimes. The best (though not always available) is a voltage and/or resistance measurement chart. This is simply a table of measurements made on an identical chassis. To be of value, the chart must be accompanied by a list of the equipment used to make the measurements, so you can use the same type of gear when you begin measuring. The reason is obvious: Because of impedance considerations, a VTVM and

VOM may deliver significantly different voltage readings when measuring identical voltages.

Voltage and/or resistance charts are often included in the repair and maintenance chapters of electronic kit instruction manuals, and in the service manuals (or on the schematic diagrams) of commercially built equipment. And they are occasionally included in project articles published in magazines and books (to save space, measurement values may be printed on the schematic diagrams, next to the key terminal points).

An alternate to a measurement table is a duplicate, working chassis that can be used as a comparison standard. Clearly, this technique works only if you've built, bought, or have access to a second chassis. Surprisingly, this often happens. For example, if one channel of a stereo amplifier may be bad, the working channel becomes your test standard.

Troubleshooting with a duplicate circuit is often more time consuming than using a table, since, unless you are familiar with the circuit design, you can't be sure which are the key readings to make. Thus, you must compare voltage and resistance measurements taken at every circuit point. And unless you are systematic as you go through the pair of chassis, it is very easy to loose your place. You may miss some points, and make repeat measurements at others. An easy way to safeguard against this is to place a pencil dot on the circuit's schematic diagram as you take a reading at the corresponding chassis point.

If neither a measurements table nor duplicate chassis is available, the only other approach is to analyze the circuit layout as best you can, and estimate the "ball park" values of correct voltage and/or resistance values. The key to doing this is to search for voltage-divider configurations of pairs of resistors, and use the appropriate formulas for gauging values. Note that you must take into account the effects of other components connected across resistors. For example, when a resistor is connected between the base and emitter terminals of a transistor, the input impedance of the transistor is effectively wired in parallel with the resistor, producing a lower net resistance for the pair. Similarly, scan the circuit carefully for coils, transformers, potentiometers and controls, and other resistors, which may be connected across the divider-pair you are looking at. Remember, a schematic diagram is often drawn with the goal of clarity in mind. It's possible that two components shown at opposite ends of the diagram are really electrically connected together. Be sure to keep track of connecting lines joining far-flung component symbols.

Actually, creating your own measurement values isn't as haphazard a business as it may at first seem. Even if your estimate is wrong, all you will lose is a bit of time, the time it takes to check out the various components around the circuit terminal that gave you the "wrong" reading when you took measurements.

A coordinated instrument troubleshooting scheme involves first locating the bad stage or portion of the circuit, and second isolating the specific bad component(s). Although there are several routes to this goal, I recommend the following: Find the problem area with voltage checks; pinpoint the specific cause with voltage, resistance, and continuity tests.

This is a logical approach, since circuit voltages are a direct guide to circuit operation (or misoperation), while resistances provide good clues to component malfunctions. We should however note again the hazard of active circuit testing (testing with the power turned on). Because the circuit must be operating in order to produce measurable voltages throughout the chassis, there is a great danger that the malfunctioning components may create circuit conditions that damage or destroy other components. For example, a short-circuited blocking capacitor may allow excessive current to flow through a transistor. Thus, as you work, keep looking for signs of component overload, such as overheating resistors and semiconductor components (these can get hot enough to cause severe burns if you touch them accidentally, so be cautious), smoking coils and transformers, and popped fuses.

Start troubleshooting by making a series of "voltage presence" tests using either your neon voltage indicator (if circuit voltages are high enough) or a VOM set to the appropriate DC voltage range (usually the range that will include the highest voltage you expect to find anywhere in the circuit—typically the power-supply output voltage). These should be fast checks—don't try to measure terminal voltages at this stage. You are verifying that voltages are present at "connected points" throughout the chassis, measured between the points and circuit ground.

What's a "connected point"? That's our term for terminals and connection points that are electrically joined to the power supply by some kind of *conductor* (a wire, a resistor, a coil, a transformer winding, a closed switch).

Compare the schematic diagram with the actual circuit carefully. Begin testing at the power supply or battery terminals and work outwards. As a rule of thumb, any connected point that is not also directly connected to ground by a wire should carry some voltage. If it doesn't, there's a good chance that the conducting component joining it to the power supply is "open circuited" (a broken connection). Or there's a possibility that some other component (of any type) that connects the point to ground may be short-circuited.

Note that these are not definite diagnoses . . . yet. It is possible that the circuit configuration is designed so that one or more connected points carry minuscule voltages. But, whenever you find a connected point voltage close to zero volts (less than $\frac{1}{10}$ volt) make a mental note to recheck the components connected to the point later on.

The next step is a bit trickier, since it requires that you analyze the schematic diagram more carefully. Search the diagram for "nonconnected points"—circuit terminals that are isolated from the power supply by transformers or capacitors. Voltage presence checks at these points should yield a zero reading. Be sure that the points you check are truly nonconnected. Remember that most other circuit components, including semiconductors, may allow current to flow. However, a definite voltage presence at a true nonconnected point *may* indicate a faulty isolating component.

Up till now, we've said nothing about voltage polarity, and we've only briefly touched upon the question of "circuit ground." Clearly, the polarity of the voltages indicated in the above test depends upon circuit design. A quick glance at the schematic diagram will tell you which lead of your voltmeter goes to the connected and unconnected points, and which lead goes to circuit ground. But what does "circuit ground" mean? We spoke about this term in Chapter 8; now we'll use a slightly different definition: *The circuit ground is the point that all circuit voltages are referenced to.*

This is a confusing way of saying something very simple. Whenever you want to make a voltage measurement of some point in a circuit, one lead of your voltmeter goes to the point in question, and the other is connected to circuit ground. In other words, by definition, circuit voltages are measured *between* the various circuit terminal points *and* circuit ground—most of the time. That "most of the time" is an annoying complication. It's possible that a circuit designer created a voltage measurement table for some project you'll troubleshoot some day by using another circuit point as the reference. If so, the fact will be stated somewhere on the chart. If no such statement is given on charts you work with, you can be sure that the circuit ground is the reference.

Incidentally, circuit ground is not necessarily the chassis. As we've said earlier in this book, some circuit designs are intended to be electrically isolated from the chassis. Thus, the circuit ground may be a buss bar or a terminal strip mounted on—but insulated from—the chassis. This will be indicated on the schematic diagram.

The next step is to begin measurement of key circuit voltages, as indicated on the voltage-reading chart. Keep in mind that circuit component tolerances and voltmeter error may cause voltage readings that differ by as much as 30 percent from table values.

Don't stop if you locate a significantly off-voltage point. There may be several more, and the more error points you find, the easier it will be to pinpoint the trouble spot. Note off-voltage points on the schematic diagram with a pencil mark—this will help you determine a pattern that may show up the circuit flaw.

It's possible that all key-point voltages will agree with the chart. This means that the trouble spot is not upsetting the circuit voltages. Don't be alarmed, since this is an equally valuable piece of information. In most circuits, only a few specific components can malfunction and not disturb circuit voltages. We'll say more about this point later.

To repeat our oft-said phrase once again: Troubleshooting isn't magic, it's detective work. Thus, off-voltage points aren't "answers"; they are clues to help you find the circuit flaw. As you'd expect, working with these clues takes experience—and luck. Happily, though, the ten guidelines listed below can often help you transform your set of clues into a pinpoint diagnosis.

1. Low power-supply voltage invariably causes low voltage readings throughout the circuit. Suspect a faulty component in the power supply *or* a short-circuited component elsewhere that could (check the schematic) partially or completely short-circuit the power-supply output to ground (search for overheated components as a starting point).

2. As a rule of thumb, an off-voltage point within a specific circuit stage is usually caused by a faulty component in that stage *or* within the stages immediately preceding and following.

3. If the output terminal of a tube or transistor is off-voltage, suspect the components and biasing network connected to its input terminal. If they are okay, check components connected to the output terminal. Lastly, suspect the tube or transistor. (Follow this rule *only* if the project has never worked correctly. If a circuit has functioned properly and then broken down, suspect the tube or transistor first.)

4. If one or more voltage readings were jittery (reading(s) fluctuated as you positioned the probe) suspect a cold-soldered or mechanically bad solder joint, *or* an intermittent failure in one of the components or sockets connected to the measurement point.

5. If the central connection between two resistances forming a voltage divider pair reads abnormally high voltage, suspect the "lower" resistance (or the wires or other components joining it to circuit ground); if abnormally low, suspect the

"upper" resistance (or the wires or conductors joining it to the power supply).

6. If two of the three terminals of a transistor read exactly the same voltage, suspect a bad transistor *or* a short-circuit among adjacent conductors or socket pins.

7. In multistage transistor amplifiers, if the output voltages of the transistors in two successive stages read low, suspect the components coupling the two stages together (especially the coupling capacitor).

8. If only one of the terminal voltages of a multiterminal coil or transformer is off-value, suspect a damaged winding.

9. If the voltages measured at both terminals of any two-terminal component (resistor, capacitor, diode) except a coil are exactly the same, suspect the component. Although certain circuit configurations develop similar (approximately) voltages on both terminals of specific components, identical voltage readings are unusual.

10. Suspect polarity errors and/or component failure if: the anode of a diode is more negative than the cathode; if the positive terminal of an electrolytic capacitor is less positive than the negative terminal; if the collector of a PNP transistor is more positive than the emitter terminal; and if the collector of an NPN transistor is less positive than the emitter terminal.

The *eleventh* guideline almost goes without saying: If you've no other clues to follow, suspect all components in the immediate vicinity (on the schematic diagram) of any off-voltage point.

The laws of troubleshooting are different than our judicial laws. All suspects are considered guilty until proven innocent. Supplying this proof is the job of the other troubleshooting techniques at your disposal.

Resistance and continuity checks are excellent for spotting open-circuited conductors, short-circuited insulators, leaky and shorted capacitors, open-circuited coils, short-circuited transformers, open and/or shorted diodes, bad switch contacts, blown fuses, bad ground connections, mechanically bad solder joints, off-value resistors and potentiometers, nonfunctioning relay contacts, poor socket connections, accidental short circuits between adjacent conductors, solder bridges on printed circuit boards, failure of insulating washers used under power semiconductor components, and a few dozen other faults that your imagination will fill in.

If your circuit is accompanied by a resistance measurement chart, by all means take point-to-point readings. These can be especially valuable when the voltage measurements don't turn up any suspects: This *may* mean that one or more components that don't have any part in establishing circuit voltages are faulty. In many circuit configurations, input and interstage transformers are connected in this manner. Check them out with resistance tests. Also suspect here are any two-terminal components wired in series with capacitors (most often resistors and coils, but occasionally diodes and switches) and any input jacks isolated from the circuit by a blocking capacitor.

Semiconductor components are best "tested" by installing a substitute part unless you own a suitable parts tester. An exception is the semiconductor diode, which can be tested with an ohmmeter (an ohmmeter circuit in a high-impedance VOM or VTVM only, to insure against damaging current flow). If the diode is functioning, the ohmmeter will read alternately high (many thousand ohms) and low resistance (less than a few hundred ohms) as the meter leads are alternately connected one way and then the other way across the diode.

Note that the installation of substitute semiconductor components should not be made until you've determined that all of the other circuit components are good. The reason is that a faulty circuit component elsewhere may have caused the first semiconductor to fail in the first place. It will do the same to the replacement device as soon as you solder it in place.

Resistance and continuity checks of specific components are best made by removing the components from the chassis, temporarily. In the case of continuity tests, the reason is that other circuit components may be damaged by the high current flowing through the device; in the case of resistance tests, other circuit components may introduce reading errors:

• The measured resistance value of a resistor wired across a diode or the terminals of a transistor will change depending upon which way the test probes are connected across the component.

• The circuit configuration may connect other resistances in parallel with the component you are measuring, producing a different net resistance value than the meter will indicate.

• A large-value electrolytic capacitor connected across the component you are measuring will act like a short circuit until current supplied by the ohmmeter circuit charges the capacitor. This can take several seconds (or even minutes) if the capacitor is very large and the ohmmeter current small (as it would be in many of the high-resistance ranges of a VTVM or TVM).

No-chart voltage checks cannot be, as we've explained, as definitive as voltage comparisons. However, they are second best, and, if made properly and combined with a bit of practical experience, they can often spotlight the circuit flaws. Moreover, you can make these checks quickly, on the spur of the moment. Here are the five guidelines to follow. Keep in mind that these rules deal with usual circuit configurations, and they may lead to misinterpretation when applied to certain common circuits:

1. Usually, DC current flows through all resistors in a circuit, except those wired in series with capactiors, and will produce voltages (in many cases very small) across the resistors. If you measure zero-volts across any resistance in the circuit, suspect nearby components that could isolate the resistance from either the power supply or circuit ground if they failed, *or* components that could short-circuit the resistor if they failed.

2. If the voltages (measured with respect to circuit ground) of any two terminals of a transistor are equal, suspect the transistor.

3. If the voltage measured between any two terminals of any component (except a tube, a coil, transformer, or switch) is zero, suspect the component *or* nearby components that could place a short circuit across the terminal pair if they failed.

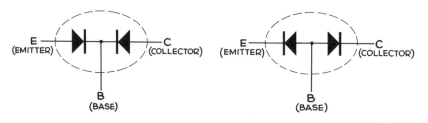

PNP TRANSISTOR MODEL

E (EMITTER)   C (COLLECTOR)

B (BASE)

NPN TRANSISTOR MODEL

E (EMITTER)   C (COLLECTOR)

B (BASE)

4. If you can measure a voltage (it will be very small) across a resistor connected in series with a capacitor, suspect the capacitor—it may be "leaking."

5. Think of each transistor in the circuit as being composed of two back-to-back diodes (see illustration). If the voltage measured between any "anode" and its "cathode" is more than ½ volt DC, suspect the transistor. Perform the same test with actual diodes in the circuit.

Glossary
Of Electronic Terms

ALTERNATING CURRENT (AC). A flow of electric current in which charged particles making up the current alternately change their direction of flow. The number of times per second that this *cycle* of direction change takes place is termed FREQUENCY, and is measured in HERTZ (or cycles-per-second). The alternating current distributed by power companies throughout the United States has a frequency of 60 Hertz. Most other countries use 50 Hertz.

AIR-DIELECTRIC CAPACITOR. *See* CAPACITOR.

ALKALINE BATTERY. *See* BATTERY.

AMMETER. An electrical instrument that measures the magnitude of an electric current flow, measured in amperes. *Also see:* PANEL METER and MILLIAMMETER.

ANTENNA. Any of a wide variety of devices designed to radiate or receive electromagnetic radiation (radio waves).

AUTOTRANSFORMER. *See* TRANSFORMER.

BATTERY. Any of a wide variety of devices made up of one or more electrochemical cells designed to produce an electric current as a result of an internal electrochemical reaction. *Primary* cells or batteries use an irreversible chemical reaction; the reacting substances are used up as the battery supplies electricity, and the battery is discarded when it can no longer supply a current. *Secondary* cells or batteries use a reversible chemical reaction; after discharge, the reacting substances within the battery can be restored to their original condition by a recharging process—usually by forcing an electric current through the discharged battery or cell.

All electrochemical cells consist of anode and cathode electrode assemblies immersed in a suitable electrolyte. By varying the materials used, many different kinds of primary and secondary batteries and cells can be created:

Alkaline cell—one electrode consists of granular metallic zinc; the other of manganese-dioxide. The electrolyte is alkaline potassium hydroxide mixed with the zinc to form a paste. The nominal output voltage of an alkaline cell is 1.5 volts DC. Alkaline cells are notable for their longer service life when compared to conventional carbon-zinc dry cells. They are primary cells.

Carbon-zinc cell (LeClanche cell)—one electrode is made of carbon; the other of zinc metal. The electrolyte is a paste made of water, zinc chloride, and ammonium chloride. This is probably the most popular type of "dry cell" (so-called because its electrolyte is a paste rather than a liquid). The familiar C and D flashlight batteries are carbon-zinc cells. They are primary cells.

Lead-acid cell—the two electrodes are made of lead peroxide and spongy lead (porous metallic lead) respectively. The electrolyte is a sulfuric acid solution. The lead-acid cell is a secondary—rechargeable—cell, and it is most widely used in batteries of three or six cells to power automotive electrical systems. When fully charged a lead-acid cell has a nominal output voltage of 2.2 volts DC.

Mercury cell—a primary cell using electrodes of zinc powder and mercuric oxide. The electrolyte is a paste of zinc oxide and potassium hydroxide. Mercury cells are noted for their ability to maintain a relatively constant output voltage (usually 1.35 or 1.4 volts DC) throughout their useful service life.

Nickel-cadmium cell—a secondary—rechargeable—cell whose electrodes are made of nickel hydroxide and cadmium, and whose electrolyte is potassium hydroxide solution. "Ni-cad" cells have become increasingly popular because their case design and construction makes them relatively leakproof; hence they are suitable for a wide variety of portable applications. Typical output voltage of a ni-cad cell is 1.2 volts DC; most cells can be recharged one or two hundred times during their life span, if allowed to discharge completely. Increased "cycling" is possible if cells are only partially discharged before recharging.

Breadboard chassis. An electronic wiring system designed to permit the rapid addition and/or removal of electronic components from an experimental circuit. Typically, components are wired to the circuit via spring-loaded clips or terminals that grip the component leads. Thus, no soldering is necessary to assemble a test circuit. The "breadboard" itself is often made of perforated chassis material to permit rapid mounting of bulky components.

Cadmium-sulphide cell. *See* Photocell.

Capacitor. An electronic circuit element consisting fundamentally of two electrically conductive surfaces (or plates) separated by a layer of electrically insulating dielectric material. The design and materials determine the capacitance (measured in Farads) of the element, and the maximum DC voltage that can be applied across the plates:

Air-dielectric capacitors have thin metal plates that sandwich a thin air film. Conventional mechanically-adjustable variable capacitors are air-dielectric units.

Ceramic capacitors consist of a thin ceramic wafer (the dielectric) upon which are deposited layers of metal film (the plates).

Electrolytic capacitors utilize aluminum foil plates separated by an ultra-thin aluminum-oxide film electrically formed on one of the plates. The nature of its design makes an electrolytic capacitor a polarized device. A DC voltage can safely be impressed across the capacitor in only one direction.

Mica capacitors are made two ways: Thin sheets of mica (the dialectric) are stacked between sheets of aluminum foil (the plates); or, thin metal films (the plates, again) are deposited on either side of a mica wafer. Mica capacitors are noted for their excellent stability—capacitance value changes little with temperature.

Mylar, paper, and plastic-film capacitors all use the same basic construction: aluminum foil plates separated by a thin mylar, paper, or plastic-film sheet, and rolled into a cylinder.

Trimmer capacitors are simply low-capacitance value adjustable capacitors that are installed in circuitry where small adjustments are needed to align or calibrate the equipment. The most commonly used trimmers are small air-dielectric or mica capacitors equipped with adjusting mechanisms.

CARBON-ZINC BATTERY. *See* BATTERY.

CARBON MICROPHONE. *See* MICROPHONE.

CARBON RESISTOR. *See* RESISTOR.

CATHODE RAY TUBE. A vacuum tube constructed so that a fast-moving beam of ELECTRONS propelled by an electron gun strikes a phosphor-coated screen and causes the phosphors to glow. By moving the beam quickly across the phosphor screen, and by modulating the beam's intensity, a C.R.T. can be used to display electronic signals (as in an OSCILLOSCOPE) or actual images (as in a television set).

CERAMIC CAPACITOR. *See* CAPACITOR.

CHASSIS. Any of a wide variety of devices used as a base or mount for electronic components forming an electronic circuit. Typically, a chassis is a boxlike design formed from aluminum or steel. However, perforated phenolic board, plastic sheet stock and boxes, and even wooden sheets and boxes are used as chassis in some applications.

CHOKE. *See* INDUCTOR.

CIRCUIT BREAKER. An electromagnetic or thermally-activated device designed to sense an overload current flow through a wire or circuit path and rapidly interrupt the current. Circuit breakers are usually resettable: pressing a button or flipping a lever restores the current flow once the circuit flaw causing the overload has been corrected. Circuit breakers are available in a wide range of preset "tripping-current" values.

COAXIAL CABLE. A two-conductor cable in which one of the conductors is a braided-wire tube that surrounds a central wire core (the second conductor). The inner wire is electrically insulated from the braided wire *shield* by plastic insulation. Because of the shielding nature of the braided wire tube, coaxial cable is used to carry audio signals (where the shield keeps out unwanted electrical noise) and radio signals (where the shield prevents the inner wire from acting like an antenna and radiating the signal).

COIL. *See* INDUCTOR.

CONDENSER. *See* CAPACITOR.

CONNECTOR. Any of a wide variety of devices used to electrically join wires, cables, or conductors to other wires, cables, or conductors, or to some part of a chassis or electronic circuit.

CRIMP-ON TERMINAL. Any of a wide variety of terminals that are mechanically joined to a wire, cable, or conductor by a crimping tool. The resulting bond between terminal and conductor is mechanically sound, and has a very low electrical resistivity (it is a highly conductive connection).

CRYSTAL. *See* PIEZOELECTRIC CRYSTAL.

CRYSTAL MICROPHONE. *See* MICROPHONE.

CURRENT. *See* ELECTRIC CURRENT.

CYCLES-PER-SECOND. *See* HERTZ.

D.C. *See* DIRECT CURRENT.

D'ARSONVAL METER. *See* PANEL METER.

DIELECTRIC. A non-conductor of electricity; an insulator.

DIODE. Any of a wide variety of devices that display the characteristic of permitting an electric current to flow in only one direction through their structures. This property is called *rectification*, and certain types of diodes which are especially built to carry large currents are known as *rectifiers*.

Germanium, selenium, and *silicon diodes* are all semiconductor diodes, and utilize a so-called semiconductor junction to achieve rectification.

Vacuum diodes are simple two-element vacuum tubes that achieve rectification by virtue of the fact that ELECTRONS will only flow from the tube's cathode to anode, and not from the anode to cathode.

DIRECT CURRENT (DC). A flow of electric current in which the charged particles making up the current flow continuously in one direction through the conductor carrying the current.

DRIVER TRANSFORMER. *See* TRANSFORMER.

DRY CELL BATTERY. *See* BATTERY.

DYNAMIC MICROPHONE. *See* MICROPHONE.

DYNAMIC SPEAKER. *See* SPEAKER.

ELECTRIC CURRENT. A flow of electrically charged particles through a suitable conducting medium. The most familiar electric current is composed of electrons moving through a metallic conductor (such as a copper wire), although other types of currents are possible, such as electrically charged ions flowing through a chemical solution.

ELECTRIC EYE. *See* PHOTOCELL.

ELECTRON. An elementary particle (a constituent part of all atoms) that possesses a negative electric charge.

ELECTROLYTIC CAPACITOR. *See* CAPACITOR.

FARAD. The unit of measurement of capacitance. In electronics, the farad is an inconveniently large measure, and the microfarad (millionth of a farad) and picofarad (millionth of a millionth of a farad) are more commonly used.

FILAMENT. The small heating element (usually made of resistance wire) inside a vacuum tube used to heat the cathode assembly. In some vacuum tubes, the filament itself serves as the cathode. In this case the heating element is coated with a substance that emits electrons when heated.

FILAMENT TRANSFORMER. *See* TRANSFORMER.

FREQUENCY. A term that denotes number of events per unit time. For example, the frequency of an ALTERNATING CURRENT is the number of times per second that the current flow reverses direction back and forth; and, the frequency of a radio wave denotes the number of oscillations per second within the wave.

FUSE. Any of a wide variety of devices that sense an overload current flowing through a conductor or circuit, and act quickly to interrupt the current flow. Typically, a fuse consists of a length of low-melting point metal wire connected in series with the circuit or conductor to be protected. An overload current heats the metal to its melting point and breaks the circuit, cutting off current flow. Fuses are available in a wide range of overload current ratings. Clearly, a fuse is a one-shot device; once it has done its job it must be discarded, and be replaced by a fresh unit.

Another class of fuses is the so-called slow-blow type. These are designed to ignore momentary overloads, and thus are used in applications where brief current surges are expected, such as in series with a motor. (When the motor starts it draws a momentarily large current.)

GENERATOR. Any of a wide variety of devices that produce electricity. All generators are really "energy converters" in that they transform some other kind of energy into electrical energy. For example, the simple DC generator used in many automobiles transforms the mechanical energy used to turn its shaft into electrical energy.

GERMANIUM DIODE. *See* DIODE.

GERMANIUM TRANSISTOR. *See* TRANSISTOR.

GIGA. The prefix used to denote one billion. For example: Two gigaohms equals two billion ohms; and five gigahertz equals five billion Hertz.

GLOW LAMP. *See* NEON BULB.

Grid. A netlike element inside a vacuum tube positioned between the cathode and plate structures.

Grommet. *See* Rubber grommet.

Ground. The reference point in a circuit from which all other circuit voltages are measured. In most electronic equipment built on or in metal chassis, the chassis is used as circuit ground.

Headphone. A device held against the ear that transforms incoming electrical signals to sound. Headphones are available with both single and double earpieces, and are primarily used for private radio listening.

Heat sink. Any of a wide variety of devices used to speed up the dissipation of heat produced by operating electronic components. In use, the heat sink and component are mechanically joined together, so that heat produced by the component can flow into the sink. The finned construction of a large heat sink permits rapid dissipation of the heat through convection.

Henry. The unit of measurement of inductance. Typically, the henry is too large a measure to be convenient in electronics, and the microhenry (millionth of a henry) and millihenry (thousandth of a henry) are used instead.

Hertz (abbreviated Hz). The unit measure of frequency. This term replaces the long-used "cycles per second." For example 10 Hz is equivalent to 10 cycles per second. *See also* Alternating current.

Inductor. An electronic circuit element consisting fundamentally of one or more coils of wire wound on an appropriate form. The size and shape of the coils, the number of turns of the coils, and the material of the form determine the inductance (measured in Henrys) of the inductor, its operating frequency range, and the maximum current that can safely flow through it. Because of their many different circuit functions, inductors are known by a variety of different names: Coil, choke, r.f. coil, r.f. choke, to list a few. *Note:* Although Transformers are actually inductors, they are so significant in their own right that we have described them in their own glossary listing.

Input transformer. *See* Transformer.

Integrated circuit. The product of a recently developed semiconductor technology that allows a complete circuit, of as many as several hundred circuit elements, to be created on a tiny chip of silicon smaller than a letter "o". I.C.'s offer many significant advantages over descrete component (conventional) circuitry besides smaller size: potentially lower cost; significantly increased reliability; improved performance.

Interstage transformer. *See* Transformer.

Isolation transformer. *See* Transformer.

Junction diode. *See* Diode.

Junction transistor. *See* Transistor.

KILO. The prefix denoting one thousand. For example: 5 kilohms equals five thousand ohms; and 10 kilovolts equals ten thousand volts.

LEAD-ACID BATTERY. *See* BATTERY.

LEAKAGE CURRENT. The undesirable current that will flow between the cabinet or exterior surfaces of an electrically powered device and any grounded object (that is electrically connected to earth ground) due to an internal insulation failure or short circuit. Leakage current is more than just an annoyance since the grounded object may be the person using the device, and the current levels involved may be hazardous.

LEVER SWITCH. *See* SWITCH.

LUG STRIP. *See* TERMINAL STRIP.

MATCHING TRANSFORMER. *See* TRANSFORMER.

MEGA. The prefix denoting one million. For example: 2 megohms equals two million ohms; and 1 megawatt equals one million watts.

MERCURY BATTERY. *See* BATTERY.

MERCURY SWITCH. *See* SWITCH.

METER. *See* PANEL METER.

MICA CAPACITOR. *See* CAPACITOR.

MICRO. The prefix denoting one-millionth. For example: 5 microamperes equals five-millionths of an ampere; and 2 microvolts equals two-millionths of a volt.

MICROFARAD. *See* FARAD.

MICROHENRY. *See* HENRY.

MICROPHONE. Any of a wide variety of transducers designed to convert sound into electrical signals. All common types of microphones utilize a diaphram, of a suitable lightweight material, that is caused to vibrate when struck by sound waves. The vibratory motion is passed on to a transducer device. Among the most familiar are:

Carbon microphone. The diaphram vibrates a layer of tight-packed carbon granules, momentarily changing the layer's electrical resistance. Thus, the flow of current through the layer is modulated in step with the sound waves.

Crystal microphone. The diaphram vibrates a piezoelectric crystal element, which produces a small modulated output voltage.

Dynamic microphone. A type of miniature electric generator in which the vibrating diaphram induces a modulated output signal in a set (one or more) of small coils.

MILLI. The prefix denoting one-thousandth. For example: 10 milliamperes equals ten-thousandths of an ampere; and 2 millivolts equals two-thousandths of a volt.

MILLIAMPERE. *See* AMPERE.

MYLAR CAPACITOR. *See* CAPACITOR.

NEON BULB. A glow discharge lamp consisting essentially of two metal electrodes contained in a glass envelope filled with neon gas at low pressure. When a sufficiently high voltage is impressed across the electrodes, the neon gas between them breaks down (ionizes) and permits a current to flow, producing a characteristic orange glow in the process.

NICKEL-CADMIUM BATTERY. *See* BATTERY.

OHM. The unit measure of electrical resistance.

OHMMETER. An electronic measuring instrument used to measure electrical resistance.

OHM'S LAW. The basic mathematical expression that defines the relationship between *voltage, current,* and *resistance* in a simple DC circuit. The expression can be written into three forms, depending on which circuit variables are known and which are unknown:

$$I = \frac{E}{R} \qquad\qquad E = I \times R \qquad\qquad R = \frac{E}{I}$$

where I = current, E = voltage, and R = resistance.

OSCILLOSCOPE. An important electronic test instrument that has the capability of pictorially displaying a varying electronic signal on a cathode-ray tube. This type of display permits a circuit designer to view different signals flowing in various legs of a complex circuit. An oscilloscope is a vital design and diagnostic tool in advanced circuit-design work.

OUTPUT TRANSFORMER. *See* TRANSFORMER.

PANEL METER. Any of a number of devices that visually display the value of an electrical quantity (volts, amperes, ohms, watts, etc.) by means of a pivoted needle-like pointer moving across a fixed scale. The instruments are constructed so that the angular distance traversed by the free end of the pointer is proportional to the quantity being measured. Thus, the numerical value on the scale that lies just below the pointer's tip corresponds to the value of the quantity.

Though several types of panel-meter movements are used (including the so-called "moving-magnet" and "iron-vane" designs), the most common to electronic experimenters is the D'Arsonval movement. Essentially, it consists of a many-turned coil of wire suspended on jeweled pivots within a magnetic structure. When direct current flows through the coil, the coil is transformed into an electromagnet—the magnetic field produced interacts with the field of the magnetic structure, and the coil rotates on its pivots. The movement's pointer is attached to the coil, so that its tip sweeps across the scale. The movement of the coil is limited by a small, coiled hairspring that winds up as the coil turns. Thus, the deflection of the pointer along the scale depends on the size of the direct current flowing through the coil—the greater the current, the greater the deflection.

Clearly, the basic D'Arsonval meter movement is a direct-current measuring instrument. However, by adding components to its input circuit it is possible to convert it into a device that will measure many other electrical quantities.

PAPER CAPACITOR. *See* CAPACITOR.

PARALLEL CIRCUIT. A circuit configuration that is arranged in such a manner that the different elements making up the circuit are connected across a source of voltage or signal. Thus, any of the circuit elements can be removed from the circuit without disconnecting the other elements. In a typical home wiring scheme, each of the wall outlets is wired in parallel with the other outlets across the incoming AC power line.

PERFORATED BOARD. *See* PHENOLIC CHASSIS BOARD.

PERMANENT MAGNET SPEAKER. *See* SPEAKER.

PHENOLIC CHASSIS BOARD. A thin board made of phenolic plastic and punched with an array of small holes that can be used as a chassis for mounting and wiring together lightweight components. Perforated boards (as these versatile boards are also called) are especially useful when assembling solid-state circuitry. The holes in the board are sized to accept PUSH-IN TERMINALS, which serve as wiring and soldering points.

PHOTOCELL. Any of a variety of devices that can respond to light in such a manner that the intensity of light striking the device's light-sensitive surface can determine a specific electrical variable such as the voltage generated by the device, the electrical resistance of the device, or the current flowing through the device.

Vacuum phototube. Light striking a photosensitive cathode structure causes electrons to be emitted, which then flow to the plate. Thus, the current flowing through the tube is proportional to the intensity of light striking the cathode.

Photovoltaic cell. Actually a light-powered electric generator. Light striking the sensitive surface causes a small DC voltage to appear across the cell's output terminals. Substantial currents can be generated by ganging several cells in parallel. Most photovoltaic cells are made of silicon or selenium. They are often called solar cells (or batteries).

Photoconductive cell. Literally a "light-controlled resistor." The electrical resistance of the cell is a function of the intensity of light striking its surface. This type of cell is often used in "electric-eye" circuits that actuate door-openers and burglar alarms. Most photoconductive cells are made of cadmium sulfide and cadmium selenide.

PICA. The prefix denoting one-millionth of a millionth. Example: 10 picafarads equals 10 one-millionth of a millionths of a farad.

PIEZOELECTRIC CRYSTAL. An electronic circuit element fabricated out of a piezoelectric material (quartz and Rochelle salt crystal, barium titanate, and lead titanate are the most commonly used materials) that has the capability of transforming a mechanical pressure applied to the element into a small electric voltage. The voltage generated is proportional to the magnitude of the mechanical pressure. In the same manner, a voltage applied across the crystal causes it to deform slightly. Piezoelectric crystal elements are widely used as transducers in microphones and phono cartridges (where they convert the motion of the diaphragm and stylus, respectively, into a corresponding output signal), and as frequency-determining elements in receivers and transmitters.

PLASTIC CAPACITOR. *See* CAPACITOR.

POTENTIOMETER. An electronic circuit element consisting fundamentally of a resistance element equipped with a movable contact tap, or "slider." Usually, the resistance element is curved into a circular shape, and the movable contact is attached to a rotatable shaft. Essentially, a potentiometer is a continuously adjustable *voltage divider:* Any fraction of the total voltage applied across the resistance element is available at the movable contact simply by placing it (by rotating the shaft) at the appropriate location along the element. Potentiometers are widely used as volume controls, and for adjusting other circuit variables.

POWER RESISTOR. *See* RESISTOR.

POWER SUPPLY. Any of a wide variety of devices designed to supply voltages and currents required to operate electronic circuitry.

POWER TRANSFORMER. *See* TRANSFORMER.

PRINTED CIRCUIT. A modern method of wiring electronic circuitry which uses a thin insulating board to which has been laminated a thin sheet of copper metal. During processing, much of the copper is chemically etched away, leaving a network of copper "lines" and "islands" that correspond to the interconnections and terminal points of the circuit. When the various required components are soldered in place on the board, the circuit is complete. Printed circuit boards (or P.C. boards, for short) are widely used in semiconductor circuit assembly, where the light weight of the components lend themselves to this kind of construction.

PUSH-IN TERMINALS. Any of a wide variety of terminals designed to mount in the small holes of a perforated phenolic chassis board. The terminals serve as wiring and soldering points.

R.F. CHOKE. *See* INDUCTOR.

R.F. COIL. *See* INDUCTOR.

RECTIFIER. *See* DIODE.

RELAY. Any of a wide variety of electromagnetically operated switches that consist fundamentally of a switch assembly that is actuated by an electromagnet. Current flowing through the relay's coil assembly produces a magnetic field that attracts an armature device, actuating the switch.

RESISTOR. An electronic circuit element consisting fundamentally of a length of electrically conductive material chosen to have some desired electrical resistance. By selecting an appropriate material, a wide range of circuit requirements can be accommodated.

Carbon resistors consist of a small chunk of carbon-composition material mounted inside an insulating sleeve. These are the most widely used resistors, and serve as general-purpose resistance elements.

Carbon film and metal film resistors consist of a conductive film deposited on an insulating core. These are usually precision-value units since resistance value can be closely controlled during manufacture.

Wire-wound resistors are made by winding a coil of resistance wire on an insulating core. Wire-wound resistors are used as precision-value units and as power resistors. (By using large-gauge wire and large central cores, large wire-wound resistors can be made to dissipate substantial amounts of power.)

ROTARY SWITCH. *See* SWITCH.

RUBBER GROMMET. Any of a variety of small rubber rings equipped with lips that mount inside holes cut in a metal chassis or cabinet. The grommet cushions the hole, and permits wires to be passed through without fear of having their insulation cut by rough metal edges.

SELENIUM PHOTOCELL. *See* PHOTOCELL.

SELENIUM RECTIFIER. *See* DIODE.

SEMICONDUCTOR. A term used generally to describe a class of chemical elements and materials that are neither good electrical insulators or electrical conductors, hence *semi*conductors. These materials are used extensively in the manufacture of transistors, diodes, silicon-controlled rectifiers, and a host of other modern components, and so, these devices are often called *semiconductor components*. Familiar semiconductors include silicon, germanium, carbon, cadmium sulfide, cadmium selenide, and galena crystal.

SEMICONDUCTOR DIODE. *See* DIODE.

SEMICONDUCTOR RECTIFIER. *See* DIODE.

SERIES CIRCUIT. A circuit configuration that is arranged in such a manner that the different elements making up the circuit are wired in line to a source of current or voltage. Thus, if any of the circuit elements are removed, current flow through the other elements is interrupted.

SIGNAL GENERATOR. Any of a variety of devices that produce signals of known waveform, frequency, and amplitude. Signal generators are extensively used during the design, calibration, alignment, and repair of electronic circuitry.

SILICON-CONTROLLED RECTIFIER. A semiconductor switching device that acts in a somewhat analogous fashion to a trap door. The application of a small electric signal to the SCR's gate electrode turns the device on, permitting a substantial current flow between the anode and cathode terminals.

SILICON DIODE. *See* DIODE.

SILICON PHOTOCELL. *See* PHOTOCELL.

SILICON RECTIFIER. *See* DIODE.

SILICON TRANSISTOR. *See* TRANSISTOR.

SLIDE SWITCH. *See* Switch.

SLOW-BLOW FUSE. *See* FUSE.

SNAP-ACTION SWITCH. *See* SWITCH.

SOLID-STATE. A term used generally to describe a class of electronic components in which current flow—and the control of current flow—take place within a solid material. This is in contrast to vacuum-tube devices, wherein current flows through an open, evacuated space inside the device. Broadly speaking, the term "solid-state" is synonymous with the term "semiconductor."

SOLAR-BATTERY. *See* PHOTOCELL.

SPAGHETTI. Thin-walled, hollow, insulating tubing (usually made of plastic or reinforced paper) that is placed over bare wires and leads to serve as insulation.

SWITCH. Any of a wide variety of devices designed to establish, interrupt, divert, transfer, or otherwise control the flow of an electric current or signal. Switches are generally classified according to two factors: (1) the number of current paths they can control; and (2) the mechanical design of the switch.

The term "poles" defines the number of current paths controlled: thus a 2-pole switch controls two individual current paths. The term "throw" defines the number of branches of the current paths made accessible as the switch is operated. Thus, a single-pole, single-throw switch is a common off-on switch; a 2-pole, double-throw switch has two current paths and has the capability of connecting each current path to one of two branches.

The commonly used mechanical designs include:

Lever switch. An external lever actuates a set of contacts.

Mercury. The contacts are bridged by a small pool of mercury whenever the switch is tilted into the operating position.

Pushbutton. A spring-loaded button actuates the contacts momentarily, deactuating them when the button is released.

Slide. The contacts are opened and closed by a movable metal bridge fastened to a sliding knob.

Snap-action. An overriding spring mechanism forces the contacts together when the switch is operated, assuring a good electrical connection. This type of switch may use lever, pushbutton, toggle, or other designs.

Rotary. The contacts are arranged in a circular configuration; a metal bridging element—called a "wiper"—that is fastened to a rotatable shaft closes the circuits as the shaft is turned.

Toggle. A spring-loaded toggle mechanism, connected to an external lever, provides a sharply defined feel of switch position as the switch is operated. Often used as on-off switches on electronic equipment.

TERMINAL STRIP. Any of a wide variety of devices designed to serve as wiring and soldering supports in and on an electronic chassis. Typically, a terminal strip consists of some type of insulating base and/or strip that carries one or more metal terminal lugs. In use, component leads and wires are connected and soldered to the lugs to electrically join them together.

TEST LEADS. Insulated conductors or cables, usually equipped with end-wired probes or connecting clips that are used to connect test equipment to a chassis or circuit under test.

THERMISTOR. A semiconductor device constructed so that its electrical resistance varies strongly with changes of the device's temperature.

TOGGLE SWITCH. *See* SWITCH.

TRANSFORMER. Any of a wide variety of devices consisting fundamentally of two or more inductor windings wound on a common core (so-called auto-transformers have one coil that is tapped in one or more places). A transformer works on a straightforward principle: When an alternating current is fed through one winding, it produces a time-varying magnetic field that is coupled through the other winding(s) by the core, inducing currents in them. By selecting the number of turns contained in each winding, and by choosing an appropriate core material, a transformer can be designed to perform a variety of functions, including transforming one AC voltage level into another, and matching a signal source having one impedance with an input having another impedance. Common types of transformers include:

Filament transformer. Transforms line-voltage AC into low-voltage AC for filaments of vacuum tubes.

Driver transformer. Used to interconnect preamplifier to power amplifier stage of audio amplifier.

Input transformer. Used to interconnect input stage of amplifier circuit with some type of transducer.

Interstage transformer. Used to interconnect stages of some type of multi-stage circuit, often an audio amplifier.

Isolation transformer. Used to isolate an AC-powered electronic device from the AC power line for safety reasons.

Matching transformer. Used to match impedances within a circuit.

Power transformer. Used to transform AC line voltage to various voltage levels required by the power supply inside some type of electronic device.

TRANSISTOR. A widely used semiconductor circuit element that functions somewhat analogously to a common water faucet. The transistor is a three-lead device, and a small electrical control signal applied to one of the leads can control the flow of electric current between the other two. Thus, the transistor can function as an amplifier: a small input signal produces a large output signal that is a replica of itself.

TRANSISTORIZED VOLTMETER. An electronic instrument, utilizing a built-in transistorized amplifier that measures AC and DC voltage. The advantage of this type of instrument when compared with a conventional panel voltmeter or VOM is that the TVM has greater input impedance, and thus loads the circuit being measured less.

TRIMMER CAPACITOR. *See* CAPACITOR.

VACUUM DIODE. *See* DIODE.

VACUUM TUBE. A widely used class of electronic devices consisting fundamentally of a series of electrodes installed within an evacuated glass or metal envelope. The cathode electrode is heated by an electrically-powered filament to a high temperature, hot enough for its treated surface to boil off electrons. These electrons are attracted to the positively charged plate electrode (which is positively charged because it is connected to the positive terminal of a high-voltage power supply). Thus, a current of electrons can flow between cathode and plate. Interspersed between cathode and plate are other electrodes called "grids." By applying signal and control voltages to these grids, the current flow can be modulated or controlled, and the vacuum tube can be used to accomplish a variety of amplification, oscillation, and control functions.

VACUUM TUBE VOLTMETER. An electronic instrument, utilizing a built-in vacuum tube amplifier, that measures AC and DC voltages (and usually resistance). The advantage of this type of instrument when compared with conventional voltmeters, or VOMs, is its much greater input impedance which loads the circuit being measured less.

VARIABLE CAPACITOR. *See* CAPACITOR.

VOLT. The unit of measurement of electromotive force, the driving pressure behind an electric current.

VOLTMETER. An electrical instrument that measures the magnitude of electronic force, or voltage. Voltmeters are calibrated in VOLTS.

WIRE-WOUND RESISTOR. *See* RESISTOR.

ZINC-CARBON BATTERY. *See* BATTERY.

Index

ELECTRONIC COMPONENTS AND THEIR SYMBOLS

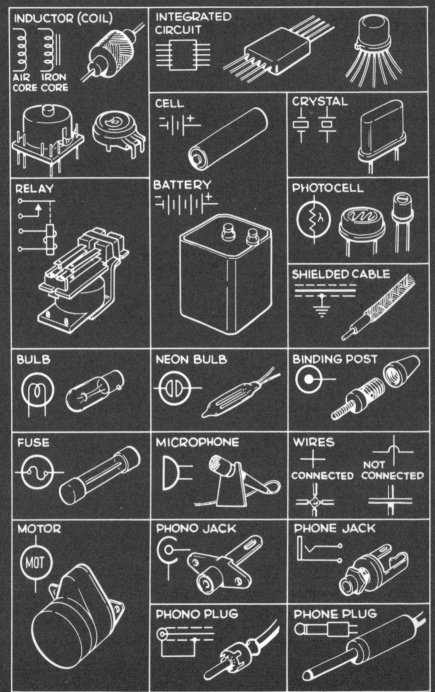

INDUCTOR (COIL)

AIR CORE   IRON CORE

INTEGRATED CIRCUIT

CELL

CRYSTAL

RELAY

BATTERY

PHOTOCELL

SHIELDED CABLE

BULB

NEON BULB

BINDING POST

FUSE

MICROPHONE

WIRES
CONNECTED   NOT CONNECTED

MOTOR

MOT

PHONO JACK

PHONE JACK

PHONO PLUG

PHONE PLUG